ELECTRIC VEHICLE
BATTERY SYSTEMS

ELECTRIC VEHICLE BATTERY SYSTEMS

Sandeep Dhameja

Newnes

Boston   Oxford   Johannesburg
Melbourne   New Delhi

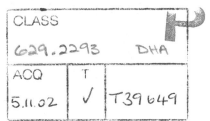

To Anju, Anita, and Aarti

Newnes is an imprint of Butterworth–Heinemann.

Copyright © 2002 by Butterworth–Heinemann

 A member of the Reed Elsevier group

Library of Congress Cataloging-in-Publication Data
Dhameja, Sandeep.
    Electric vehicle battery systems / Sandeep Dhameja.
        p.   cm.
    Includes bibliographical references and index.
    ISBN 0-7506-9916-7
    1. Automobiles, Electric—Batteries.   I. Title.
    TL220 .D49 2001
    629.22'93—dc21   2001030855

British Library Cataloguing-in-Publication Data
A catalogue record for this book is available from the British Library.

The publisher offers special discounts on bulk orders of this book.
For information, please contact:

Manager of Special Sales
Butterworth–Heinemann
225 Wildwood Avenue
Woburn, MA 01801-2041
Tel: 781-904-2500
Fax: 781-904-2620

For information on all Newnes publications available, contact our World Wide Web home page at: http://www.newnespress.com.

10 9 8 7 6 5 4 3 2 1

Printed in the United States of America

TABLE OF CONTENTS

ACKNOWLEDGMENTS   ix

1   ELECTRIC VEHICLE BATTERIES   1
Electric Vehicle Operation   2
Battery Basics   4
Introduction to Electric Vehicle Batteries   4
Fuel Cell Technology   14
Choice of a Battery Type for Electric Vehicles   18

2   ELECTRIC VEHICLE BATTERY EFFICIENCY   23
Effects of VRLA Battery Formation on Electric Vehicle Performance   23
Regenerative Braking   24
Electric Vehicle Body and Frame   24
Fluids, Lubricants, and Coolants   25
Effects of Current Density on Battery Formation   25
Effects of Excessive Heat on Battery Cycle Life   35
Battery Storage   35
The Lithium-ion Battery   39
Traction Battery Pack Design   41

3   ELECTRIC VEHICLE BATTERY CAPACITY   43
Battery Capacity   43
The Temperature Dependence of Battery Capacity   44
State of Charge of a VRLA Battery   46
Capacity Discharge Testing of VRLA Batteries   51
Battery Capacity Recovery   53
Definition of NiMH Battery Capacity   54
Li-ion Battery Capacity   58
Battery Capacity Tests   60
Energy Balances for the Electric Vehicle   64

4 ELECTRIC VEHICLE BATTERY CHARGING 69
Charging a Single VRLA Battery 69
Charge Completion of a Single VRLA Battery 69
Temperature Compensation During Battery Charging 72
Charging NiMH Batteries 74
Rate of Charge Effect on Charge Acceptance Efficiency of Traction
    Battery Packs 74
Environmental Influences on Charging 80
Charging Methods for NiMH Batteries 81
Charging Technology 87
Battery Pack Corrective Actions 91

5 ELECTRIC VEHICLE BATTERY FAST CHARGING 95
The Fast Charging Process 95
Fast Charging Strategies 98
The Fast Charger Configuration 101
Using Equalizing/Leveling Chargers 105
Inductive Charging—Making Recharging Easier 111
Range Testing of Electric Vehicles Using Fast Charging 113
Electric Vehicle Speedometer Calibration 114

6 ELECTRIC VEHICLE BATTERY DISCHARGING 115
Definition of VRLA Battery Capacity 117
Definition of NiMH Battery Capacity 119
Discharge Capacity Behavior 123
Discharge Characteristics of Li-ion Battery 127
Discharge of an Electric Vehicle Battery Pack 128
Cold-Weather Impact on Electric Vehicle Battery Discharge 130

7 ELECTRIC VEHICLE BATTERY PERFORMANCE 133
The Battery Performance Management System 133
BPMS Thermal Management System 137
The BPMS Charging Control 141
High-Voltage Cabling and Disconnects 148
Safety in Battery Design 150
Battery Pack Safety—Electrolyte Spillage and Electric Shock 153
Charging Technology 155
Electrical Insulation Breakdown Detection 157
Electrical Vehicle Component Tests 157
Building Standards 159
Ventilation 159

8 TESTING AND COMPUTER-BASED
MODELING OF ELECTRIC VEHICLE BATTERIES   161
Testing Electric Vehicle Batteries   163
Accelerated Reliability Testing of Electric Vehicles   167
Battery Cycle Life versus Peak Power and Rest Period   171
Safety Requirements for Electric Vehicle Batteries   188

APPENDIX A: FUEL CELL PROCESSING
TECHNOLOGY FOR TRANSPORTATION
APPLICATIONS: STATUS AND PROSPECTS   191
APPENDIX B: VEHICLE BATTERY CHARGING
CHECKLIST/LOG   205
APPENDIX C: DAY 1/2/3 RANGE AND CHARGE TEST LOG   207
APPENDIX D: SPEEDOMETER CALIBRATION
TEST DATA LOG   209
APPENDIX E: ELECTRIC VEHICLE PERFORMANCE TEST
SUMMARY   211

BIBLIOGRAPHY   215

INDEX   221

ACKNOWLEDGMENTS

This book would not have been possible without the help and support of a number of people, and I would like to express my gratitude to all of them. Robert Jacobs for giving me the opportunity to work with Daimler-Chrysler. Min Sway-Tin and Jim Cerano for giving me the opportunity to be part of the EPIC electric vehicle (EV) development team. The late Sandy Cox for providing with a better understanding of the behavior of the EV batteries, which led to the further investigation of the formation characteristics of the EV batteries and also motivated me to author a practical understanding of EV battery design.

I would also like to thank my family for their patience and encouragement—my mother, Anju, and father, Rajesh Kumar, my wife, Anita, and my sister Aarti. I would also like to thank my father in particular for providing me with the idea for the book design.

I would like to thank Otilio Gonzalez for reviewing my manuscript preparation contract and all the members of the EPIC EV team for providing me the motivation to author this book.

Finally, I would also like to thank Carrie Wagner at Newnes/Butterworth–Heinemann, for making this manuscript a publication success.

1 ELECTRIC VEHICLE BATTERIES

Road vehicles emit significant air-borne pollution, including 18% of America's suspended particulates, 27% of the volatile organic compounds, 28% of Pb, 32% of nitrogen oxides, and 62% of CO. Vehicles also release 25% of America's energy-related CO_2, the principle greenhouse gas. World pollution numbers continue to grow even more rapidly as millions of people gain access to public and personal transportation.

Electrification of our energy economy and the rise of automotive transportation are two of the most significant technological revolutions of the twentieth century. Exemplifying this massive change in the lifestyle due to growth in fossil energy supplies. From negligible energy markets in the 1900, electrical generation now accounts for 34% of the primary energy consumption in the United States, while transportation consumes 27% of the energy supply. Increased fossil fuel use has financed energy expansions: coal and natural gas provide more than 65% of the energy used to generate the nation's electricity, while refined crude oil fuels virtually all the 250 million vehicles now cruising the U.S. roadways. Renewable energy, however, provides less than 2% of the energy used in either market.

The electricity and transportation energy revolution of the 1900s has affected several different and large non-overlapping markets. Electricity is used extensively in the commercial, industrial and residential sectors, but it barely supplies an iota of energy to the transportation markets. On the other hand oil contributes only 3% of the energy input for electricity. Oil usage for the purpose of transportation contributes to merely 3% of the energy input for electricity. Oil use for transportation is large and growing. More than two-thirds of the oil consumption in the United States is used for transportation purposes, mostly for cars, trucks, and buses. With aircraft attributing to 14% of the oil consumption, ships and locomotives consume the remaining 5%. Since the United States relies on oil imports, the oil use for transportation sector has surpassed total domestic oil production every year since 1986.

The present rate of reliance and consumption of fossil fuels for electrification or transportation is 100,000 times faster than the rate at which they are being created by natural forces. As the readily exploited fuels continue to be consumed, the fossil fuels are becoming more costly and difficult to extract. In order to transform the demands on the development of energy systems based on renewable resources, it is important to find an alternative to fossil fuels. Little progress has been made in using electricity generated from a centralized power grid for transportation purposes. In 1900, the number of electric cars outnumbered the gasoline cars by almost a factor of two. In addition to being less polluting, the electric cars in 1900 were silent machines. As favorites of the urban social elite, the electric cars were the cars of choice as they did not require the difficult and rather dangerous handcrank starters. This led to the development of electric vehicles (EVs) by more than 100 EV manufacturers.

However, the weight of these vehicles, long recharging time, and poor durability of electric barriers reduced the ability of electric cars to gain a long-term market presence. One pound of gasoline contained a chemical energy equivalent of 100 pounds of Pb-acid batteries. Refueling the car with gasoline required only minutes, supplies of gasoline seemed to be limitless, and the long distance delivery of goods and passengers was relatively cheap and easy. This led to the virtual disappearance of electric cars by 1920.

ELECTRIC VEHICLE OPERATION

The operation of an EV is similar to that of an internal combustion vehicle. An ignition key or numeric keypad is used to power up the vehicle's instrumentation panels and electronic control module (ECM). A gearshift placed in Drive or Reverse engages the vehicle. When the brake pedal is released, the vehicle may creep in a fashion similar to an internal combustion vehicle. When the driver pushes the accelerator pedal, a signal is sent to the ECM, which in turn applies a current and voltage from the battery system to the electric motor that is proportional to the degree to which the accelerator pedal is depressed. The motor in turn applies torque to the EV wheels. Because power/torque curves for electric motors are much broader than those for internal combustion (IC) engines, the acceleration of an EV can be much quicker. Most EVs have a built-in feature called regenerative braking, which comes into play when the accelerator pedal is released or the brake pedal is applied.

This feature captures the vehicle's kinetic energy and routes it through the ECM to the battery pack. Regenerative braking mimics the deceleration effects of an IC engine.

An appealing quality of EVs is that they operate very quietly. For the most part, the handling and operation of commercial EVs is comparable to their internal combustion counterparts.

Electric Vehicle Components

The major components of the EV are an electric motor, an ECM, a traction battery, a battery management system, a smart battery charger, a cabling system, a regenerative braking system, a vehicle body, a frame, EV fluids for cooling, braking, etc., and lubricants. It is important to look at the individual functions of each of these components and how they integrate to operate the vehicle.

Electronic Drive Systems

An EV is propelled by an electric motor. The traction motor is in turn controlled by the engine controller or an electronic control module. Electric motors may be understood through the principles of electromagnetism and physics. In simple terms, an electrical conductor carrying current in the presence of a magnetic field experiences a force (torque) that is proportional to the product of the current and the strength of the magnetic field. Conversely, a conductor that is moved through a magnetic field experiences an induced current. In an electric propulsion system, the electronic control module regulates the amount of current and voltage that the electric motor receives. Operating voltages can be as high as 360V or higher. The controller takes a signal from the vehicle's accelerator pedal and controls the electric energy provided to the motor, causing the torque to turn the wheels.

There are two major types of electric drive systems: alternating current (AC) and direct current (DC). In the past, DC motors were commonly used for variable-speed applications. Because of recent advances in high-power electronics, however, AC motors are now more widely used for these applications. DC motors are typically easier to control and are less expensive, but they are often larger and heavier than AC motors. At the same time, AC motors and controllers usually have a higher efficiency over a large operational range, but, due to complex electronics, the ECMs are more expensive. Today, both AC and DC technologies can be found in commercial automobiles.

BATTERY BASICS

A battery cell consists of five major components: (1) electrodes—anode and cathode; (2) separators; (3) terminals; (4) electrolyte; and (5) a case or enclosure. Battery cells are grouped together into a single mechanical and electrical unit called a battery module. These modules are electrically connected to form a battery pack, which powers the electronic drive systems.

There are two terminals per battery, one negative and one positive. The electrolyte can be a liquid, gel, or solid material. Traditional batteries, such as lead-acid (Pb-acid), nickel-cadmium (NiCd), and others have used a liquid electrolyte. This electrolyte may either be acidic or alkaline, depending on the type of battery. In many of the advanced batteries under development today for EV applications, the electrolyte is a gel, paste, or resin. Examples of these battery types are advanced sealed Pb-acid, NiMH, and Lithium (Li)-ion batteries. Lithium-polymer batteries, presently under development, have a solid electrolyte. In the most basic terms, a battery is an electrochemical cell in which an electric potential (voltage) is generated at the battery terminals by a difference in potential between the positive and negative electrodes. When an electrical load such as a motor is connected to the battery terminals, an electric circuit is completed, and current is passed through the motor, generating the torque. Outside the battery, current flows from the positive terminal, through the motor, and returns to the negative terminal. As the process continues, the battery delivers its stored energy from a charged to a discharged state. If the electrical load is replaced by an external power source that reverses the flow of the current through the battery, the battery can be charged. This process is used to reform the electrodes to their original chemical state, or full charge.

INTRODUCTION TO ELECTRIC VEHICLE BATTERIES

In the early part of 1900s, the EV design could not compete with the plethora of inventions for the internal combustion engine. The speed and range of the internal combustion engines made them an efficient solution for transportation. By the middle of the 1900s, discussions about the impending oil supplies, the growing demands of fossil fuels began to rekindle the inventions of alternate energy systems and discovery of alternate energy sources. By the mid-1970s, oil shortages led to aggressive development of EV programs. However, a temporarily stable oil supply thereafter and a rather slow advancement in alternate energy technology for traction batteries once again impeded EV development.

In the 1990s, concerns both over the worldwide growth of demand for fossil fuels for transportation, namely petroleum and the reduction of vehicle emissions has once again intensified EV development. This in turn has led to advances in research and development of traction batteries for EVs.

The U.S. Department of Energy (DOE) has formed the U.S. Advanced Battery Consortium (USABC) to accelerate the development of advanced batteries for use in EV design. The Consortium is a government-industry partnership between DOE and the three largest automobile manufacturers—Daimler-Chrysler, Ford, and General Motors—and the Electric Power Research Institute (EPRI). The USABC has established battery performance goals intended to make EVs competitive with conventional IC engine vehicles in performance, price, and range. The path of technological development for EV batteries will emphasize advanced Pb-acid, NiMH batteries, Li-ion, and lithium-polymer batteries. Daimler-Chrysler, Ford, and General Motors will initially use Pb-acid batteries. Honda and Toyota will produce vehicles that use nickel metal-hydride batteries, while Nissan will demonstrate vehicles using Li-ion batteries.

Some of the salient features of the traction battery for EVs are:

- High-energy density can be attained with one charge to provide a long range or mileage
- The high-energy density makes it possible to attain stable power with deep discharge characteristics to allow for acceleration and ascending power capability of the EV
- Long cycle life with maintenance free and high safety mechanisms built into the battery
- Wide acceptance as a recyclable battery from the environmental standpoint

For over a century, the flooded lead-acid batteries have been the standard source of energy for power applications, including traction, backup or standby power systems. With significant advances in research, over the last decade, the development of the valve regulated lead-acid (VRLA) battery has provided for an alternative to the flooded lead-acid battery designs. As the user demand for VRLA batteries continues to grow for traction battery applications, more energy density per unit area is being demanded. It is thus important to understand the benefits and limitations of VRLA.

VRLA battery technology for traction applications arose from demands for a "no maintenance" battery requiring minimal attention. Especially for maintaining distilled water levels to prevent drying of cells and safe operation in battery packs in EV applications. However, it can be argued that to the present day, a true "no maintenance" battery does not exist. Rather the term "low maintenance" battery is a more suitable term.

Two types of VRLA traction batteries are available commercially, the absorbed glass mat (AGM) battery and the gel technology battery. Each of the battery designs is similar to the common flooded lead-acid battery.

The Pb-Acid Battery

A flooded or wet battery is one that requires maintenance by periodic replenishment of distilled water. The water is added into each cell of the battery through the vent cap. Even today, some large uninterruptible power supply applications use flooded lead-acid batteries as a backup solution. Although they have large service lives of up to 20 years, they have been known to be operational for a longer time (up to 40 years for a Lucent Technologies round cell).

The design of flooded lead-acid battery comprises negative plates made of lead (or a lead alloy) sandwiched between positive plates made of lead (or a lead alloy) with calcium or antimony as an additive. The insulator (termed as a separator) is a microporous material that allows the chemical reaction to take place while preventing the electrodes from shorting, owing to contact.

The negative and the positive plates are pasted with an active material—lead oxide (PbO_2) and sometimes lead sulphate ($PbSO_4$). The active material provides a large surface area for storing electrochemical energy. Each positive plate is welded together and attached to a terminal post (+). Using the same welding each negative plate is welded together and attached to a terminal post (−). The plate assembly is placed into a polypropylene casing. The cover with a vent cap/flame arrestor and hydrometer hole is fitted onto the container assembly. The container assembly and the cover plate are glued to form a leak-proof seal. The container is filled with an electrolyte solution of specific gravity 1.215. The electrolyte solution is a combination of sulphuric acid (H_2SO_4) and distilled water.

Upon charging or application of an electric current, the flooded lead-acid battery undergoes an electrochemical reaction. This creates the cell's potential or voltage. Based on the principle of electrochemistry, two dissimilar metals (positive and negative plates) have a potential difference (cell voltage). Upon assembly of the plates, a float charge is placed on the battery to maintain a charge or polarization of the plates.

During the charge phase, water in the electrolyte solution is broken down by electrolysis. Oxygen evolves at the positive plates and hydrogen evolves at the negative plates. The evolution of hydrogen and oxygen results in up to 30% recombination. A higher battery efficiency means that no watering is required, sharply reducing the maintenance cost compared to the flooded lead-acid battery. It is the recombination factor that improves the VRLA battery efficiency. In the VRLA battery, the efficiency is 95 to 99%. Special ventilation and acid containment requirements are minimal with VRLA batteries. This allows batteries to be colocated alongside electronics. The two types of VRLA batteries are the absorbed glass mat (AGM) based battery and the gel technology battery.

As the name suggests, the AGM based VRLA battery is much like the flooded battery because it uses standard plates. In addition, it has a higher specific gravity of the electrolyte solution. The glass mat is used to absorb and contain the free electrolyte, essentially acting like a sponge. The AGM allows for exchange of oxygen between the plates also termed as recombination. At the same time the glass mat provides electrical separation or insulation between the two negative and positive plates of the battery.

The thickness of the glass mat determines the degree of absorption of the electrolyte solution. The greater the ability to store electrolyte, the lower is the probability of the cell dry out. This prevents the shorting of the plates. The AGM battery's safety vent or flame arrestor is the second difference from the flooded Pb-acid battery design. The valve or flame arrestor prevents the release of oxygen during normal battery operation. It maintains the internal battery pressure for recombination of the electrolyte. In addition, it acts as a safety device in preventing sparks and arcs from entering the cell (much like flooded lead-acid batteries). And, in case of excessive gas pressure build-up, the vent acts as a relief.

The second VRLA battery is based on gel technology. This battery also uses plates and electrolyte as in the flooded Pb-acid batteries. A pure form of silica is added to the electrolyte solution forming an acidic gel. As the gel dries out, cracks are formed. The cracks, when seen through a clear casing, appear identical to a shaken bowl of gelatin. These cracks in the acidic gel are useful and allow diffusion of oxygen between the positive and the negative plates. Thus making it a recombinant gel technology. The acidic gel in a higher fluid form is referred to as Prelyte and enhances the oxygen diffusion thus improving the battery life.

The gel technology, like the AGM battery, is also fitted with a vent or flame arrestor to maintain the internal battery pressure, preventing the release of hydrogen and oxygen during abnormal operation.

The specific gravity of the batteries in comparison is between 1.215 for the flooded Pb-acid and 1.300 for the VRLA battery. The volume of the available electrolyte is an important factor in determining the battery performance. Thus the flooded battery with a lower Ahr rating exhibits long-rate performance than the larger VRLA battery since they have a larger acid reservoir.

In addition, AGM battery designs have the highest performance because they have the lowest internal resistance and higher gravity electrolyte (1.300) in comparison with their counterparts. End-voltage ratings and Ahr measurements are insufficient factors to base a conclusion about flooded batteries with respect to VRLA designs. It is important to consider battery ventilation, space requirements, acid containment, economic practicality as other factors affecting battery selection.

Table 1–1 indicates the costs associated with battery maintenance; installation service is based on a $60 per hour cost and the IEEE recommendations.

The NiMH Battery

The NiMH battery is considered to be a successor to the long-time market dominator—the Nickel Cadmium (Ni-Cd) battery system. These cells have been in existence since the turn of the century. The Ni-Cd battery system started with a modest beginning, but with significant

Table 1–1 Costs associated with battery maintenance.

Feature	Flooded ($)	AGM ($)	Gel ($)	Modular AGM ($)
Battery Price	20,000	24,000	20,000	19,000
Rack Price	2,200	2,200	2,200	—
Spill Containment	1,700	—	—	—
Installation	5,000	5,000	5,000	3,600
Ventilation	2,000	—	—	—
20-Year Maintenance	14,400–45,000	7,200–38,500	7,200–35,000	7,200–30,000
Initial Installation Cost	30,000	31,000	27,000	22,000
Annual Cost	2,500	2,000	2,000	1,500

advances in the last four decades since the 1950s, the specific capacity of the batteries has improved fourfold. A strong growth of the rechargeable battery consumer appliance market for laptop computers, mobile phones, and camcorders pushed the battery performance requirements—particularly service output duration—even further. This factor, along with environmental concerns, has accelerated the development of the alternate NiMH system. Since its inception in the early to mid-1980s, the market share of the rechargeable NiMH battery has grown to 35% and the capacity, particularly the high-load capability, has been improving dramatically.

The scientific publications and patent literature provide an extensive number of reports regarding the different aspects of NiMH batteries, including chemistry and hydrogen storage properties of cathode materials. However, it is important to understand design criteria that optimize performance and extend the cycle life of NiMH batteries.

AB_5 ($LaNi_5$) and AB_2 (TiN_2) alloy compounds have been studied as part of NiMH battery design. Both these alloys have almost similar hydrogen storage capacities, approximately 1.5% by weight. The theoretical maximum hydrogen storage capacities of AB_2 alloys is slightly higher, 2% by weight than the maximum of 1.6% by weight for AB_5 alloys. The higher AB_2 hydrogen storage capacity by weight can be exploited only if the battery size is made larger. This becomes an undesirable factor for compact EV battery designs. The basic concept of the NiMH battery cathode results from research of metallic alloys that can capture (and release) hydrogen in volumes up to a thousand times of their own. The cathode mainly consists of a compressed mass of fine metal particles. The much smaller hydrogen atom, easily absorbed into the interstices of a bimetallic cathode is known to expand up to 24 volume percent. The hydride electrode has capacity density of up to $1,800 \, mAh/cm^3$.

Thus for the smaller size NiMH battery, the higher energy density for AB_5 alloys, about 8–$8.5 \, g/cm^3$ compared to relatively lower energy density for AB_2 alloys, about 5–$7 \, g/cm^3$ results in a battery with comparable energy density.

The conventional, although not cost-effective processing method for manufacturing the AB_5 battery materials includes:

Step 1: Melting and rapidly cooling of large metals ingots
Step 2: Extensive heat treatment to eliminate microscopic compositional inhomogeneities
Step 3: Breaking down the large metal ingots into smaller pieces by the hydriding and dehydriding process
Step 4: Grinding of the annealed ingots pieces into fine powders

This four-step manufacturing process is the key-limiting factor to widespread commercialization of NiMH batteries. This process can be eliminated and replaced by a single step using rapid solidification processing of AB_5 powders using high-pressure gas atomization. The H_2 gas absorption and desorption behavior of the high-pressure gas atomization processed alloy is also significantly improved with the annealing of the powder.

The Li-ion Battery

Li-ion batteries are the third type most likely to be commercialized for EV applications. Because lithium is the metal with the highest negative potential and lowest atomic weight, batteries using lithium have the greatest potential for attaining the technological breakthrough that will provide EVs with the greatest performance characteristics in terms of acceleration and range. Unfortunately, lithium metal, on its own, is highly reactive with air and with most liquid electrolytes. To avoid the problems associated with metal lithium, lithium intercalated graphitic carbons (Li_xC) are used and show good potential for high performance, while maintaining cell safety.

During a Li-ion battery's discharge, lithium ions (Li^+) are released from the anode and travel through an organic electrolyte toward the cathode. Organic electrolytes (i.e., nonaqueous) are stable against the reduction by lithium. Oxidation at the cathode is required as lithium reacts chemically with the water of aqueous electrolytes. When the lithium ions reach the cathode, they are quickly incorporated into the cathode material. This process is easily reversible. Because of the quick reversibility of the lithium ions, lithium-ion batteries can charge and discharge faster than Pb-acid and NiMH batteries. In addition, Li-ion batteries produce the same amount of energy as NiMH cells, but they are typically 40% smaller and weigh half as much. This allows for twice as many batteries to be used in an EV, thus doubling the amount of energy storage and increasing the vehicle's range.

There are various types of materials under evaluation for use in Li-ion batteries. Generally, the anode materials being examined are various forms of carbon, particularly graphite and hydrogen-containing carbon materials. Three types of oxides of transition are being evaluated for the cathode: cobalt, nickel, and manganese. Initial battery developments are utilizing cobalt oxide, which is technically preferred to either nickel or manganese oxides. However, cobalt oxide is the costliest of the three, with nickel substantially less expensive and manganese being the least expensive.

In Li-ion batteries in which cobalt oxide cathodes are used, the cathodes are currently manufactured from an aluminum foil with a cobalt-oxide coating. The anodes are manufactured using a thin copper sheet coated with carbon materials. The sheets are layered with a plastic separator, then rolled up like a jellyroll and placed inside a steel container filled with a liquid electrolyte containing lithium hexafluoro-phosphate. These batteries have an open circuit voltage (OCV) of approximately 4.1 V at full charge.

In addition to their potential for high-specific energy, Li-ion batteries also have an outstanding potential for long life. Under normal operation, there are few structural changes of the anodes and cathodes by the intercalation and removal of the smaller lithium ions. Additionally, the high voltage and conventional design of Li-ion batteries hold the promise of low battery cost, especially when cobalt is replaced by manganese.

Overcharging of Li-ion batteries, as with Pb-acid and NiMH batteries, must be carefully controlled to prevent battery damage in the form of electrode or electrolyte decomposition. Because the electrolyte in a lithium-ion battery is nonaqueous, the gassing associated with water dissolution is eliminated. The development of advanced battery management systems is a key to ensuring that lithium-ion batteries operate safely, during normal operation as well as in the event of vehicle accidents. As with Pb-acid and NiMH batteries, Li-ion battery charging systems must be capable of working with the battery management systems to ensure that overcharging does not occur. The solid-state rechargeable Li-ion battery offers higher energy per unit weight and volume. In addition, the Li-ion is an environmentally friendly battery in comparison with nickel-based batteries, which use NiMH battery chemistry.

Commercialization of these Li-ion batteries was achieved fairly quickly in the 1960s and 1970s. The development of lithium rechargeable batteries was much slower than their NiMH and Pb-acid counterparts due to battery cell failure caused by lithium dendrite formation and an increased reaction of high-area lithium powders formed by cycling. To overcome the battery failure, alternative solutions to metallic lithium were proposed. An alternative material based on carbon involves an innovative design, called the rocking-chair or shuttlecock, in which the lithium ions shuttle between the anode and the cathode. During the discharge process, lithium ions move from the anode to the cathode. During the charge process, lithium ions move from the cathode to the anode. The voltage of the lithiated anode is close to that of lithium metal (approximately +10 mV), and, hence, the cell voltage is not reduced significantly.

Lithium ions shuttle between the anode and the cathode with minimal or no deposition of the metallic lithium on the anode surface as in the case of lithium metal rechargeable batteries. Thus making the Li-ion batteries safe for use.

Solid-state Li-ion batteries offer several advantages over their liquid electrolyte counterparts. Although the liquid Li-ion batteries have been around for several years, the solid-state Li-ion battery introduced in 1995 into the commercial market is substantially superior. Energy densities exceed 100 Whr/kg and 200 Whr/L. The operating temperature of these batteries is also wide, from –20°C to 60°C.

Sony Corporation incorporated the rocking-chair concept into the design of Li-ion cells for commercial applications. Ever since Sony Energytec, Inc., introduced the Li-ion battery in 1991, the development efforts have been burgeoning. Sony Corporation announced a production increase to 15 million batteries per month in 1997. The polymer gel electrolyte development was motivated by the safety concerns. Sony developed the fire-retardant electrolyte that forms a skin of carbon molecules. The skin prevents evaporation of the organic solvents and isolates the electrolyte from combustion-supporting oxygen. Boosting the production to 30 million batteries per month in the years since 1997.

During the charging process, the Li-ion cell anode equation is represented as:

$$Li_xC_6 + xLi^+ + xe^- \rightarrow LiC_6$$

And the Li-ion cell cathode equation is represented as,

$$LiCoO_2 \rightarrow xLi + xe^- + Li(1 - x)CoO_2$$

During the discharging process, the Li-ion cell anode equation is represented as,

$$LiC_6 \rightarrow Li_xC_6 + xLi^+ + xe^-$$

And the Li-ion cell cathode equation is represented as,

$$xLi^+ + xe^- + Li(1 - x)CoO_2 \rightarrow LiCoO_2$$

The Sony Corp. Li-ion cell is composed of the lithiated carbon anode, a Li_xCoO_2 cathode and a nonaqueous electrolyte. Other battery manufacturers have followed with variations of the same basic cell chemistry for EV applications.

Table 1–2   Development of Li-ion battery systems.

Year	Cathode	Anode	Electrolyte	Battery System
1980–1990	$LiWO_2$	$LiCoO_2$, $LiNiO_2$	Polymer	Li/MoO_2, $LiVO_x$
1990–2000	LiC_6	$LiMn_2O_4$		$C/LiMn_2O_4$

The VARTA Li-metal oxide/carbon system is known under the Li-ion or Swing system. Both the electrodes reversibly intercalate resulting in the release of Lithium without changing their host structure. The Li-ion battery operates at room temperature. Owing to its high cell voltage level, the battery requires an organic electrolyte.

The first Li-ion cells for EV applications were based on the $LiCoO_2$ (lithium-cobalt-oxide) cathode and demonstrated a capacity of 30 Ahr. Once detailed analyses results for the anode and $LiCoO_2$, $LiNiO_2$, and $LiMnO_4$ based cathodes were available, battery manufacturers decided to focus on the development of the lithium-manganese spinel. In addition to the 30 Ahr cell, two other cell types based on the $LiMn_2O_4$ were developed by VARTA. All Li-ion cells have a prismatic steel casing and stacked electrode configuration. Since the performance of the large prismatic cells with a specific energy greater than 100 Whr/kg and the specific power greater than 200 Whr/kg meet the requirements of EV battery applications, intensive research efforts for low cost positive electrode materials have led to significant electrode material developments. By synthesizing a special lithium manganese oxide spinel structure with a specific capacity almost identical to the cobalt oxide spinel, 60 Ahr battery cells are now available and capable of providing a specific energy of 115 Whr/kg.

Table 1–2 summarizes the progress made between 1980 and 2000 in the development of Li-based battery systems.

The Li-Polymer Battery

Lithium-polymer batteries are the fourth most likely type of battery to be commercialized for EV applications. The discovery of nonmetallic solids capable of conducting ions has allowed for the development of these batteries. Lithium-polymer batteries have anodes made of either lithium or carbon intercalated with lithium. One candidate cathode under evaluation contains vanadium oxide (V_6O_{13}). This particular battery chemistry has one of the greatest potentials for the highest specific energy and power. Unfortunately, design challenges associated with kinetics of the battery electrodes, the ability of the cathode and anode to absorb and release lithium ions, has resulted in lower specific power and limited cycle life for lithium-polymer batteries.

The current collector for lithium-polymer batteries is typically made of either copper or aluminum foil surrounded by a low thermal conductivity material such as polyurethane. The battery case is made of polypropylene, reinforced polypropylene, or polystyrene.

Lithium-polymer batteries are considered solid-state batteries since their electrolyte is a solid. The most common polymer electrolyte is polyethylene oxide complexed with an appropriate electrolyte salt. The polymers can conduct ions at temperatures above about 60°C (140°F), allowing for the replacement of flammable liquid electrolytes by polymers of high molecular weight. Since the conductivity of these polymers is low, the batteries must be constructed in thin films ranging from 50 to 200 μm thick. There is, however, a great safety advantage to this type of battery construction. Because the battery is solid-state by design, the materials will not flow together and electrolyte will not leak out in case there is a rupture in the battery case during an EV accident. Because the lithium is intercalated into carbon anodes, the lithium is in ionic form and is less reactive than pure lithium metal. Another major advantage of this type of battery construction is that a lithium-polymer battery can be formed in any size or shape, allowing vehicle manufacturers considerable flexibility in the manner in which the battery is incorporated into future vehicle designs.

FUEL CELL TECHNOLOGY

The oil crisis in 1973 led to the development of the alternative automotive power sources. This development of alternative power sources prompted EV for urban transportation. During this period, the primary concern was to gain independence from foreign oil sources. The two primary commercially available battery types were the Pb-acid and the NiCd batteries. This prompted research into the development of fuel cells. In the case of the battery, chemical energy is stored in the electrode, while in the case of the fuel cell, the energy is stored outside the electrodes. Thus there is no physical limit to the amount of fuel stored. This is analogous to the gasoline cars with internal combustion engines. Renewable energy-based hydrogen vehicles used in place of conventional and diesel-fueled internal combustion engines will reduce automotive air pollution significantly.

Dating back to the developments in 1839, Sir William Graves first demonstrated the fuel cell principle. Since 1987, the DOE has awarded several development contracts, including the development of small urban bus systems powered by methanol-fueled phosphoric acid fuel cells (PAFC). In addition, the developments include a 50-kW proton exchange membrane fuel cell (PEMFC) propulsion system with an onboard methanol reformer and direct hydrogen-fueled PEMFC systems for development of mid-size EVs.

Graves based the discovery of the principle of thermodynamic reversibility of the electrolysis of water. The reversible electrochemical reaction for the electrolysis of water is expressed by the equation:

$$\text{Water} + \text{electricity} \leftrightarrow 2H_2 + O_2$$

Electric current flow was detected through the external conductors when supplying hydrogen and oxygen to the two electrodes of the electrolysis cell. When more than one fuel cell was connected in series, an electric shock was felt, which led to the representation of the above equation as:

$$2H_2 + O_2 \leftrightarrow 2H_2O + \text{electricity}$$

Hydrogen gas is supplied to the anode and reacts electrochemically at the electrode surface to form protons and electrons. These electrons travel through the electrode and connecting conductors to an electric load, such as a motor, and over to the fuel cell's cathode. At the cathode, the electrons react with the oxygen and the previously produced protons to form water. The presence of platinum (Pt) increases the speed of the chemical reaction to produce electric current. The anodic and cathodic reactions may be expressed as:

$$\text{Anode: } H_2 \rightarrow [M1]2H^+ + 2e^-$$

$$\text{Cathode: } O_2 + 4H^+ + 4e^- \rightarrow 2H_2O$$

The different types of fuel cell technologies include five major fuel cell designs, each described by the conducting electrolyte in the cell. The anodic and the cathodic reactions for the fuel cells do tend to differ. In both the alkaline and the acidic fuel cells, the electrolyte's conducting species are protons and hydroxide ions—the products of water's electrolytic dissociation.

Table 1–3 provides a comparison between the types of electrolyte, their operating temperature range, and efficiency.

Table 1–3   Comparison between electrolytes fuel-cell electrolytes.

Type	Electrolyte	Temperature (°C)	Features
Alkaline	KOH (OH⁻)	60–120	High efficiency
PEMFC	Polymer electrolyte (H⁺)	20–120	High-power density
PAFC	Phosphoric acid (H⁺)	160–220	Limited efficiency
MCFC	Molten Carbonate (CO_3^-)	550–650	Complex control
SOFC	Solid doped Zr-oxide (O⁻)	850–1,000	Stationary, power generation

The fuel for operating a fuel cell is not limited to hydrogen, and the overall electrochemical reaction is given by:

$$Fuel + oxidant \rightarrow H_2O + other\ products + electricity$$

Water and electricity are the only products of the hydrogen-fueled fuel cells. For most fuel cells, the thermodynamic efficiency is about 90%. The efficiency of the fuel cell is defined as the ratio of the free energy and the enthalpy of the electrochemical reaction.

Identical to the losses in the thermal engine, the electrochemical engine loses efficiency due to the over-potential of the anodic and cathodic reactions, internal cell resistance, and the mass transfer problems. An additional benefit of the fuel cell is that it has a higher efficiency at partial loads, while the IC engine is much more efficient at full power.

The open circuit voltage (OCV), or the equilibrium voltage, of the fuel cell is 1.25 V, under no load condition. As soon as the current flows through the cell with a load connected to the terminals, fuel cell voltage drops, and the efficiency of the fuel cell drops.

The initial and steep voltage drop represents the over-potential of the electrode reactions. This is the voltage drop due to the current withdrawn from the fuel cell. It is also the sum of the cathodic and the anodic over-potentials. Defined in simple terms, it is the voltage needed to overcome the potential barrier of the oxygen and the hydrogen electrode reactions. The next gradual potential drop is the voltage drop at high current density due to inadequate mass transfer of the reacting species of the electrodes. The advances in the fuel-cell technology thus far demonstrate that the current density of 0.8–1.2 A/cm^2 is possible from a single fuel cell within the range of 0.55–0.75 V.

Several single fuel cells are connected in series or parallel to form a fuel cell battery stack of desired voltage, current, and power. One or more of the fuel cell stacks are constructed with pumps, humidifiers, and gas filters to form a fuel cell system. The air compressor pressurizes the air, increasing the supply of oxygen, while the humidifier provides the PEM membrane an aqueous environment to conduct the protons through the electrolyte.

While the energy consumption of the ancillary equipment is low, practical efficiency of the system will be even lower than that derived from the polarization curve. Estimates of the total efficiency loss of the fuel cell subsystem are in the range of 40 to 50%, thus reducing the practical efficiency of 50 to 60%. This is still higher than the practical thermal efficiencies of 25 to 35% for the heat engine.

Considerable amounts of platinum (Pt) catalyst are required to achieve current levels to meet traction requirements. On the contrary fuel cell stacks applications must reduce the Pt catalyst content and improve stack operation with air.

In 1990, $4\,mg/cm^2$ of Pt was required at a cost of \$75/kW. In 1993, a tenfold reduction of Pt was achieved, reducing the cost to \$8/kW. New developments have led to Pt reduction down to $0.15\,mg/cm^2$. Thus reducing costs of the Pt electrode even further to \$3/kW.

Improvements in the membrane assembly design (MEA) have resulted in assembly of fuel cell stacks with higher power densities. In 1989, the power density of a typical stack was 0.085 kW/L, which increased to 0.14 kW/L in 1991. In 1993, the fuel cell stack power density doubled to 0.29 kW/L and then reached 0.72 kW/L at the end of 1994. Fuel cell stack power density reached 1 kW/L in 1998. Recent demonstrations of fuel cell stack power density have exceeded 1.5 kW/L.

Usually, power density of the cell subsystem is 30 to 35% of the stack's power density. This equates to a system power density of 0.3 to 0.35 kW/L. Most thermal combustion engines have power densities of approximately 1 kW/L. This power density demonstrates that significant amounts of developments are required for gaining parity with IC engines. In terms of costs, the IC engine powertrain costs range between \$20/kW to \$30/kW.

In the case of the direct PEMFC based engine, significant amounts of onboard hydrogen storage are required. In the chemical equivalent, the hydrogen is stored in rechargeable metal hydride, or in a hydride-based compound that releases hydrogen upon contact with water. Physically, hydrogen can be stored as compressed gas, cryogenically cooled liquid, or through adsorption on a surface of the membrane.

The automotive and process industry has considerable experience with onboard compressed natural gas (CNG) storage. CNG storage requires 25 MPa of pressure to contain the volume of CNG necessary for 560 km of driving. Hydrogen is a much lighter gas requiring a much higher pressure. The compressed hydrogen tank size required to contain

6.8 kg of hydrogen for a 3-L, 1,500-kg vehicle with a driving range of 560 km is 340 L at 25 MPa, and 160 L at 52 MPa. A typical gas tank volume for such a vehicle is 70 L. Thus the limited energy storage capacity of hydrogen and the lack of an infrastructure to supply it makes it necessary to develop a process to extract hydrogen from gasoline.

The Daimler-Chrysler experimental fuel-processing technology converts gasoline into hydrogen, carbon dioxide (CO_2), and water (H_2O) in a multistage chemical reaction process. The five stage processing components consist of the following:

Fuel Vaporizer   By applying heat, liquid gasoline is converted to gases to ensure low pollution. The vaporized gas during combustion passes on to the next stage.

Partial Oxidation (POX) Reactor   Vaporized fuel is combined with some air in a Partial Oxidation reactor, producing H_2 and CO.

Water-Gas Shift   Steam as the catalyst converts most of the CO to harmless CO_2 and additional H_2. Since CO is harmful to both, excessive inhalation and the fuel cell. Thus the concentration of CO must be reduced to less than 10 ppm.

Preferential Oxidation (PROX)   In the PROX, the injected air reacts with the remaining CO. With steam as the catalyst the preferential oxidation process results in production of CO_2 and hydrogen-rich gases.

Fuel Cell Stack   The hydrogen gas, combined with air, produces electricity to move the vehicle with virtually no pollution—with the emission of water vapor.

The greatest challenges facing the changes in transportation are the lack of understanding of the broad range of consequences of environmental pollution and reliance on IC engine based transportation. In addition, the lack of confidence in the alternate fuel technology is the key deterrent of commercialization of the alternative fuel based technology transportation.

The increase in the hydrogen program expenditures over the past decade can be summarized in Table 1–4. The increase in the annual expenditure demonstrates a significant promise in the fuel cell based vehicles for both commercial and domestic passenger vehicles.

CHOICE OF A BATTERY TYPE FOR ELECTRIC VEHICLES

VRLA battery designs operate successfully in partially closed environments. They do not require as much floor space as their flooded lead-

Table 1–4   Spending between 1992 and 2000 for the hydrogen program.

Fiscal Year	Expenditure (million $)
1992	1.4
1993	3.8
1994	9.5
1995	10
1996	14.5
1997	15
1998	18
1999	20
2000	25

acid type counterparts. In addition, they certainly do not require as much maintenance. As they continue to decrease in size, they are improving in energy density and cost.

NiMH batteries are also termed environmentally friendly and continue to improve both in energy density and cost.

Li-ion batteries are capable of storing up to three times more energy per unit weight and volume than the conventional Pb-acid and NiMH batteries. This is approximately three-times voltage level of 3.5 V. Because of the high-energy characteristics, Li-ion batteries find widespread applications including aerospace, EV, and hybrid EV designs. However, the scaling of the consumer Li-ion cells is necessary.

While evaluating battery suitability for unique applications, it is important to understand a variety of battery characteristics, including the energy/power relationship (Ragone Plot), battery and cell impedance as a function of temperature, pulse discharge capability as a function of both temperature and load, and battery charge/discharge characteristics.

The self-discharge rate of the solid-state Li-ion battery is fairly low—5% of the capacity per month, compared to the 15% for the VRLA battery and 25% for NiMH battery. There is no memory effect in the solid-state Li-ion battery as is the case in the NiMH and the VRLA battery. The battery cycle life is superior to the NiMH and VRLA batteries. In the case of the NiMH battery, the cycle life typically drops to 80% of the rated capacity after 500 cycles at the C-rate (one hour charge followed by a one hour discharge). Solid-state Li-ion batteries can achieve more than 1,200 cycles before reaching 80% of their rated capacity.

Figure 1–1   Life cycle of a Li-ion EV cell.

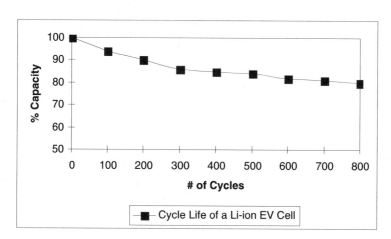

— Cycle Life of a Li-ion EV Cell

Table 1–5   Developing Li-ion battery chemistry and characteristics.

Battery Capacity	5 hours C/5	Ahr	65
Energy Density	5 hours	Whr/l	270
Specific Energy	5 hours	W/kg	115
Power Density	30 sec 80% DOD	W/l	435
Specific Power	30 sec 80% DOD	W/kg	180
Cycle Life	80% of Capacity		700
Rate Capability	Cap@C/1/Cap@C/5	%	80
Charge Time		Hr	4–5

Li-ion batteries, particularly solid-state batteries are efficient at charge-discharge rates other than the C-rate. In addition, the liquid Li-ion batteries are not suited for use in EVs owing to safety reasons while the solid-state batteries are well suited for high-rate applications.

Solid-state Li-ion batteries allow for the development of virtually any size batteries. In addition, the batteries can be stacked into efficient multicell configurations. From a cost perspective, the solid-state Li-ion battery uses a relatively inexpensive metal oxide that is fabricated in sheet form to allow inexpensive battery production. Electrodes, electrolyte, and foil packaging—all on a continuous-feed roll—are sandwiched together into finished batteries in one integrated process. By comparison, the liquid Li-ion battery cells require a cumbersome winding and canning process. Thus in comparison, solid Li-ion batteries will be easily mass-produced at less than a $1 per Whr. The NiMH battery, after years of improvements, is being produced at approximately the same cost, $1 per Whr.

The solid Li-ion batteries are safer to produce than the liquid Li-ion batteries because the solid polymer electrolyte is both nonvolatile and leak-proof. There is no chance for the Li-ion battery cell to be breached leading to an electrolyte leak.

Future developments of solid-state Li-ion batteries go into full-scale commercial production. Efforts to enhance the energy density and rate capability of the next generation batteries are already underway. The U.S. Advanced Battery Consortium (USABC) is funding research to improve the ion conductivity of solid-state electrolytes.

A large number of characteristics of the Li-ion battery are favorable for EV applications. These include:

- High gravimetric and volumetric energy densities
- Ambient temperature operation
- Long life cycle (See Figure 1–1)
- Good pulse power density

The $LiMn_2O_4$ oxide based Li-ion battery is:

- Considerably cheaper
- More environmentally benign
- Less toxic than $LiCoO_2$ and $LiNiO_2$ based batteries

In addition, the $LiMn_2O_4$ based EV batteries demonstrate a cycle life between 700 to 1,000 cycles before the capacity of the battery drops down to 80% of its initial capacity under room temperature conditions. Table 1–5 summarizes the developing Li-ion battery chemistry and characteristics.

The next generation design efforts include:

- to further extend the battery service life to 10 years
- to cut the battery costs significantly

2 ELECTRIC VEHICLE BATTERY EFFICIENCY

To develop a practical electric vehicle (EV), it is essential to understand the behavior of a set of batteries in a pack (when they are connected in a series or parallel configuration).

Figure 2–1 illustrates some of the factors that influence the battery efficiency of a traction battery (battery efficiency η is defined as the ratio of the energy output of a battery to the input energy required to restore the initial state of charge under specified conditions of temperature, current rate, and final voltage). Formation time, discharge rate, frequency of the charge/discharge of batteries, temperature of charging and discharging, etc., are some of the factors that affect the battery efficiency.

EFFECTS OF VRLA BATTERY FORMATION ON ELECTRIC VEHICLE PERFORMANCE

The formation process of batteries converts unformed paste masses to the charged state, which is stored as energy. In the case of lead-acid batteries, this involves the formation of lead dioxide (PbO_2) with lead on the positive plate and sponge lead on the negative plate. This process of conversion of the unformed paste to an active mass is carried out in the presence of sulfuric acid. At the end of formation, the battery gains its final strength and the active mass exhibits a porous structure with a large internal surface.

However, several factors affect the formation of the batteries, including electrolyte temperature, concentration of the forming electrolyte, and current density during the formation cycles. These external factors singularly affect the battery efficiency η and performance of the EV. For example, a typical 85 Ahr valve regulated lead acid battery requires 8 to 12 cycles to form up to 60% of its useful rated capacity (capable of providing 60 to 65 Ahr), up to 17 to 20 cycles to form up to 65% of its useful rated capacity (capable of providing 75 Ahr) and up to 36 to 40 cycles

Figure 2–1   Factors affecting the battery efficiency η.

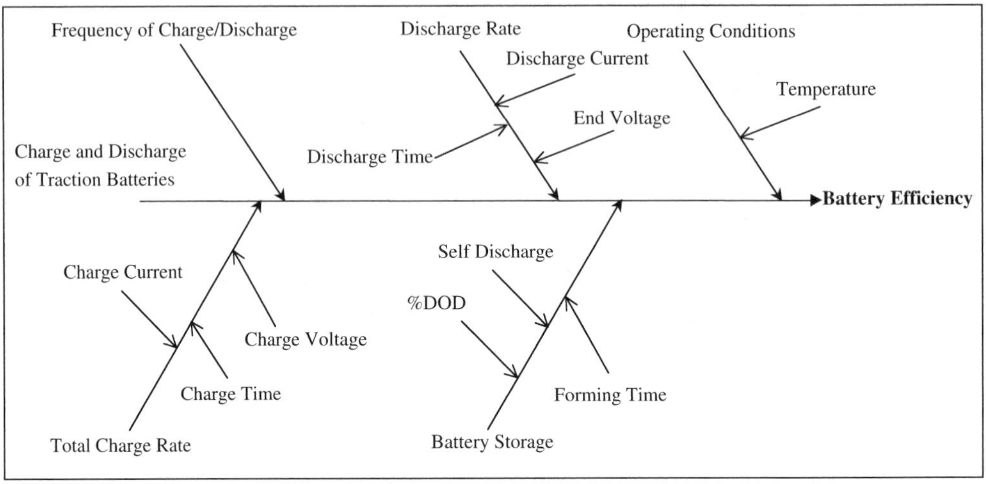

to form up to 100% of its useful rated capacity (capable of providing 85 to 95 Ahr).

REGENERATIVE BRAKING

Another key aspect of EVs is that they can recapture kinetic energy and store it as electrical energy through a process called regenerative braking. As discussed earlier, electric current is drawn from the battery system and applied, under the management of the electronic control module, to the motor. Inside the motor, this current is passed through a magnetic field, which results in a torque that is used to turn the vehicle's wheels. This process can be reversed during braking of the vehicle. Effectively, the electronic control module converts the motor to a generator. The momentum stored in the moving vehicle is used to pass the conductors in the motor through a magnetic field, which creates a current that is then directed by the electronic control module back to the battery system where it is stored for future use. Regenerative braking systems can increase the driving range of EVs by 10 to 15%.

ELECTRIC VEHICLE BODY AND FRAME

Many of the materials used in EV bodies and frames are the same as those used in internal combustion vehicles. Steel is commonly used in those EVs that have internal combustion vehicle chassis. Some of the EV designs use aluminum frames. Many EV body panels are manufactured of the same materials as those used in internal combustion vehicles, such as plastics, steel, and composites. Some EVs use magnesium alloys in certain components to reduce vehicle weight. Magnesium alloys have been used in various internal combustion vehicle components as well, such as differential covers, engine blocks, and wheels.

FLUIDS, LUBRICANTS, AND COOLANTS

As with internal combustion vehicles, there are a number of fluids used in EVs. For example, hydraulic fluid is used in the brake and power steering systems, while coolants are used in the motors and electronic control modules of some vehicles. Refrigerants and lubricants are used in the air conditioning compressor; and lubricant is also used in the motor, differential, and gearbox of the drive train. In most cases, these lubricants and coolants are the same as those used in internal combustion vehicles.

Antifreeze coolants may be employed in a number of passenger and heavy-duty vehicles to cool the motor, the controller unit, and the battery pack. The amount of fluid varies, with 2 to 3 gallons being standard.

EVs do not store large quantities of flammable liquids. Many manufacturers use electric-powered heat pumps to heat and cool the passenger compartment. These pumps are effective only when the ambient temperature is greater than 40°F. Because some vehicles are used in locations where the ambient temperature drops below 40°F, other heating systems are employed. Heaters fired by compressed natural gas, diesel, propane, or kerosene may be optional equipment employing less than a few gallons.

EFFECTS OF CURRENT DENSITY ON BATTERY FORMATION

If the current density of the batteries is changed during their formation, either in the charge or discharge process or in both processes, the capacity of the batteries is affected. With rising currents and falling temperature, the capacity of the battery decreases.

However, the capacity of the battery remains stable during cycling and maintains sufficient porosity for an exceptionally high battery discharge rate capability.

Effects of Current Density on VRLA Battery Formation

With lower discharge currents the capacity depends less on the electrical load and tends to its maximum available-value. After the battery is discharged at low temperatures (between 25°C and −25°C) during formation, a larger discharge capacity can be achieved at elevated temperatures. Thus indicating that the maximum capacity value is reached only when the battery undergoes low current discharges. In addition, room temperature capacity of the battery can be obtained at modestly low temperatures only if the battery is discharged with smaller loads.

Effects of Current Density on NiMH Battery Formation

The electrical capacity of NiMH batteries exceeds that of nickel-cadmium (NiCd) batteries by about 40%. The NiMH battery is environmentally friendly and contains no toxic cadmium. It provides high-energy storage with reduced weight and volume.

During the charging process, hydrogen is generated at the nickel hydroxide anode. The bivalent nickel hydroxide is oxidized to trivalent state by electron removal. At the same time, protons are discharged at the cathode's metal/electrolyte boundary layer in the form of neutral hydrogen atoms. The hydrogen atoms enter the bulk of the metal alloy where they are stored as metal hydride.

The absorption and desorption of hydrogen results in expansion and subsequent contraction of the cathode. The large volume expansions of the cathode induce structural defects. Metal hydride electrode structure changes occur as a result of battery cycling. Large particles of the electrode become an agglomeration of 1 to 12 particles after cycling. The large electrode particles are reduced in size within 1 to 5 battery cycles. This reduction in the particle size continues through the battery life to over 1,000 cycles. At this point the electrode particle size is reduced to approximately 1μ to 3μ and leads to a consequent reduction of bimetallic cathode particle size.

The reduction of particle size causes an increased diffusion of the metal at the surface of the cathode. The metal readily oxidizes to form a thick layer of hydroxide, which acts as an inhibitor to the oxygen recombination cycle. As the reaction rate is increased, the oxygen, and even hydrogen, is now generated faster than it can be recombined. Thus, the hydrogen storage capacity of the NiMH battery is also reduced. This phenomenon results in the increase of the electrode temperature and battery pressure. In order to maintain a pressure equilibrium within the

Figure 2–2   Variation of battery cell voltage during early formation cycles.

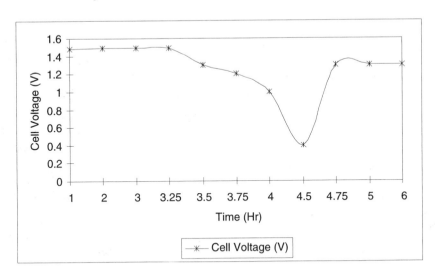

cell, the safety vents release the gases, resulting in the loss of useful water vapor.

The development of corrosion-resistant metal hydride (MH) alloys provides for a design of NiMH batteries that can perform at internal temperatures above 45°C. Some of the commercially available AB_5 battery alloys can result in venting of sealed battery cells in less than 150 cycles. Increase of the electrode surface area by reduction of both the alloy particle size and grain size decreases the hydrogen diffusion path resulting in good capacity and a long cycle life.

During the early formation cycles, a discharge current at 7.0 A corresponds to a C/3 rate. The useful output capacity of a 1.5 V cell is approximately 65% (17 Ahr). Figure 2–2 illustrates the variation of the battery cell voltage during the early battery formation cycles.

Most of the electrode mechanical degradation occurs at very high and very low states of charge. If the hydride is not fully utilized during battery cycling, the degree of embrittlement will be lower. Maintaining the electrode at full charge or full discharge is not as mechanically-aggressive as cycling to high depths of discharge.

Effects of Electrode Oxidation on NiMH Battery Formation

Surface properties of the MH affect the electrochemical properties of the NiMH battery. Oxidation of the NiMH electrode, occurs after two to three battery charge-discharge cycles, decreases charging efficiency of the battery by reduction of the voltage that induces release of hydrogen. The electrode conductive sites are covered by oxide resulting in the reduction of catalytic activity and gas recombination.

Cracking of the electrode surface area occurs during the first 5 to 20 NiMH battery cycles. X-ray photoelectron spectroscopy demonstrates that exposure to the potassium hydroxide (KOH) electrolyte does affect the surface area. The surface roughness increases due to electrolyte exposure.

A steady state oxide layer surface during an X-ray photoelectron spectroscopy demonstrates that there is no distinct oxide layer with a sharp interface. Furthermore, since the NiMH electrode can have various oxidation states, the oxide is not uniform even at a defined depth. Oxidation of the electrode also reduces the hydrogen storage capacity. One possible mechanism involves decreased charge efficiency, although the discharge efficiency is also reduced due to cell polarization. Excessive oxidation, can result in an overall decrease in conductivity or an isolation of powder particles.

Extensive battery oxidation occurs upon exposure to the electrolyte. Oxidation causes lower specific capacity, decreased rate capability, decreased charging efficiency, slower gas recombination, and ultimately, NiMH cell failure.

Effects of Elevated Temperatures on VRLA Battery Formation

Formation temperature affects the structure of the sponge lead (active material) of Pb-acid traction batteries. At elevated temperatures, the active material (paste) undergoes recrystallization, affecting the structure of the active material to produce coarse lead sulfate ($PbSO_4$) crystals. Formation of $PbSO_4$ crystals results in formation of a barrier on the active material. As a result, batteries require a higher current to store energy (penetrate the barrier layer and are difficult to charge). In addition, the $PbSO_4$ crystals continue to form even after the forming current is applied to the batteries.

A decrease in the forming temperature is accompanied by a decrease in the percentage of lead dioxide (PbO_2) for temperatures between 28°F and 158°F. Table 2–1 summarizes the effects of formation on a lead acid battery cell at a C/5 discharge rate at varying temperatures. Thus, an EV with a fully charged battery pack at 77°F will provide only 60% (approximately) of its useful rated capacity under cold temperature conditions. This factor will limit the driving range of the EV significantly.

Table 2–1   Effect of temperature on VRLA cell capacity (assuming electrolyte density = 1.340 kg/ltr).

Temperature (°F)	Capacity (%)
–10.4 (frozen)	9
23	57
33.8	68
50	82
59	87
68	94
77	100

Effects of Temperature on NiMH Battery Formation

During formation, the kinetic properties of the cathode are decisive in determining the cycle life of the NiMH battery. During charging, the NiMH battery is limited by the temperature rise of the cells resulting from exothermic hydride formation. At elevated temperatures, the corrosion of the MH battery alloy also increases substantially.

NiMH batteries are designed to withstand overcharge to maximize battery capacity and simplify charge control systems. During a battery overcharge, the negative electrode is particularly susceptible to corrosion owing to oxygen evolved at the Ni positive electrode. The oxygen is evolved during chemical recombination at the MH electrode.

Additional MH electrode capacity is designed into the NiMH battery. This excess capacity is designed into the NiMH cell to prevent evolution of hydrogen under overcharge condition. Thus conserving the escape of the electrolyte through the safety vent. Lost electrolyte results in increased battery resistance leading to the loss of capacity and failure of the cell.

The absorption resistance—ability of the cathode to accept charge in the form of neutral protons increases as depth of discharge (%DOD) increases. Furthermore, useful AC impedance measurements have shown that the Diffusion Time Constant (strongly dependant on the length of the diffusion path) increases with increasing %DOD.

Activation and Formation of a NiMH Battery

It is important to distinguish the activated surface of a NiMH battery from the electrode surface during manufacturing. The surface of the electrode can be characterized as a thin, dense, passivated oxide film. The oxide is formed due to the exposure of the NiMH battery electrode to the atmosphere or high temperature exposure to impurities or during the processing steps.

In most cases, the *fabricated* surface of the NiMH electrode is not suitable for electrochemical operation. Owing to a poor charge acceptance, the electrode surface requires an etch treatment. Etch treatment provides an electrochemically active surface, which also allows charge acceptance. During the charging of a NiMH battery, the absorbed hydrogen expands the metal lattice of the electrode, which cracks and in turn creates a new surface area. This electrode surface is large and oxide-free. Surface analysis of the electrode demonstrates that the oxygen gas recombination occurs rapidly at the outer surface of the NiMH battery electrode.

End of Formation of a VRLA Battery

A number of criteria are useful to determine the end of formation of a VRLA battery including:

- Cell voltage within a module becomes constant; the values depend upon the temperature and internal resistance (specific gravity)
- Cells gas uniformly and strongly
- Temperature of the cells within a module rises steeply towards the end of the formation in the equalization phase if the current is not reduced

During the charging process, hydrogen is generated at the nickel hydroxide anode. The bivalent nickel hydroxide is oxidized to trivalent state by electron removal. At the same time, protons are discharged at the cathode's metal/electrolyte boundary layer in the form of neutral hydrogen atoms. The hydrogen atoms enter the bulk of the metal alloy where they are stored as metal hydride.

The absorption and desorption of hydrogen results in expansion and subsequent contraction of the cathode. The large volume expansions of the cathode induce structural defects and consequent reduction of bimetallic cathode particle size. The reduction of particle size causes an increased diffusion of the metal at the surface of the cathode. The metal readily oxidizes to form a thick layer of hydroxide. This layer acts as an inhibitor to the oxygen recombination cycle. The oxygen, and even hydrogen, is now generated faster than it can be recombined. In order to maintain pressure equilibrium within the cell, the safety vents release the gases, which results in the loss of water vapor.

Thus, while considering the reaction mechanism of a NiMH battery, the charging product of the hydrogen-absorbing pressure must be controlled. The upper limit of the pressure is typically 20 atmosphere (atm.) at 70°C. This value is determined using a selected hydrogen-absorbing alloy with an equilibrium absorbing pressure and an adequate capacity balance between the positive and the negative electrodes. The discharge product of the hydrogen-absorbing alloy electrode is water. The water diffuses into the electrolyte solution. The electrolyte concentration is maintained under an upper limit of 35 wt.% independent of both the battery scale and shape. A concentration higher than 35 wt.% will result in degradation of the positive electrode.

Failure Modes of VRLA

Most battery manufacturers prefer not to discuss or talk about the possible failure modes of battery failure. Rather they are interested in talking about the length and the dependable life of the batteries. However, it remains to be a fact that batteries do fail.

In VRLA lead-acid batteries, grid corrosion is the leading cause of battery failure. It is the corrosion of the positive grid that limits the life expectancy of the battery. As the battery cycles and ages, the grid gradually begins to deteriorate losing its conductivity. In addition, the adhesion of the active paste materials also decreases. The material lost from the plates, lead/sulphur salt falls to the bottom of the cell container and gradually build up. The build up eventually grows to form a conductive path leading to shorting off the battery. This leads to reduction of the cell capacity. Analogous to deposits in a water pipe that restricts the flow of water through a pipe, corrosion on the plate restricts the flow of current in the battery.

Another common failure of VRLA batteries is grid growth. During the life of the battery, the plates tend to expand dimensionally. This change in size exerts pressure on the casing, vents, and the terminal posts. Since the weakest link in the battery assembly remains to be the terminal posts, the expansion forces them upwards to crack the casing-to-cover seal thus leading to premature battery failure.

As the battery ages, the grid deteriorates inwardly and loses its ability to support the current. The corrosion is analogous to a 4 American Wire Gauge (AWG) wire being reduced down to 12 AWG wire. Eventually, the electrical path to the terminal post is lost. In order to prevent premature battery failure owing to grid corrosion, the traction battery manufacturers must provide grid thickness and cycle life data for EV applications. The thicker the battery grid is, the longer the usable life of the battery is. Furthermore, the thicker grid provides better mechanical support for the active material and reduces the porosity of the plate material.

Thicker grids for VRLA batteries are manufactured using bottom-pour casting resulting in low porosity that dramatically reduces the rate of grid corrosion. However, users of VRLA batteries have witnessed premature battery failures owing to the defects formed during production, application environment designs, charging control mechanisms, and manufacturing techniques.

After years of development and perfecting design attributes, manufacturers have found one area of VRLA battery design that they have not been able to control—the self-discharge tendency of the negative plates of the battery. In addition, the capacity performance of the battery tends to "fall off" prematurely. This "fall off" in battery capacity happens in ideal float operating conditions because of discharged negative plates.

Another common failure is due to discharge of the negative plate over a period of time. This occurs owing to the high recombination efficiency of the battery. As the battery continues to age, the recombination efficiency continues to improve (greater that 98% in some cases). The oxygen released at the positive plate, during the charging process, is absorbed by the negative plate. This reduces the spongy lead active material and prevents the negative plates from reaching the fully charged state.

The failure modes due to battery operating conditions and their effects are illustrated in Figure 2–3.

In some studies, it was believed that as long as enough negative active material existed, the negative plates could remain in a slightly discharged state without substantial loss of battery performance. Water loss in the VRLA battery is mainly attributed to the loss of hydrogen that is given off by the corrosion process at the positive grid. The water loss occurs when some of the oxygen created at the positive plate does not diffuse into the negative plate. Instead the oxygen is depleted by the corrosion of the positive grid.

The resulting depolarization or loss of charge of the negative plate allows self-discharge. Self-discharge is a chemical reaction that leads to the discharge of the negative plate. The reaction removes the sulphate and lowers the electrolyte specific gravity. This lowering in the specific gravity in turn leads to lowering the open circuit voltage (OCV).

In some cases a boost charge may be applied to the discharged negative plate. The boost charge is also referred to as an equalization charge for battery packs. Typically, this boost charge requires to be closely mon-

Figure 2–3   Battery failure modes due to operating conditions.

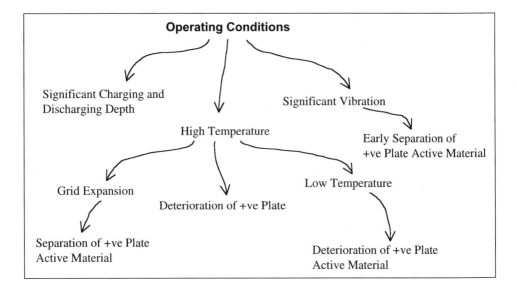

itored and varies from 2.45 V/cell to 2.5 V/cell. However, this equalization charge results in gassing and generates heat for a short time.

In VRLA batteries, the oxygen given off is at the top of the charge cycle. The oxygen off the positive plates causes heating and gassing of the VRLA battery. Since all the oxygen off the positive plates is either recombined at the negative plates or used to corrode the positive grid. The hydrogen is either recombined at the negative plates or used to corrode the positive grid. The hydrogen that is given off as a by-product of the positive plate corrosion is vented out of the cell in periodic intervals. In addition, hydrogen is also vented out due to battery self-discharge at the negative plates. The combined loss of hydrogen is the main reason for the battery water loss.

The new design of VRLA battery includes a catalyst in the cell. The catalyst recombines the oxygen and the hydrogen to recover water lost during electrolysis. In VRLA battery cells, the catalyst attracts and recombines some of the oxygen that would normally be diffusing to the negative plates. This oxygen recombines with the hydrogen present in the headspace of the battery.

As a result, the negative plate is slightly polarized and obtains a full charge. A slight amount of hydrogen evolved during the process, recombines at the catalyst surface. Here are some of the benefits of the catalyst:

Consistent performance throughout the battery life   The polarization on the negative plates is maintained at full state of charge. As a result, the battery capacity performance does not fall off prematurely.

Longer battery life   The polarization on the positive plates is lowered, reducing the battery corrosion rate at the positive grid.

Reduction in float current   Since the plates are polarized, a lower float current is required to maintain full charge on the battery cells.

Reduction in the battery thermal runaway potential   Since there is a lower current through the cell, there is less internal heating and thus a lower thermal runaway potential.

Reduction in the battery impedance   Since the battery functions at an improved state of charge, the cell's impedance is lowered.

Reduction in the battery water loss   Less venting causes less water loss from the battery.

Under high temperature conditions, the float charge increases. The higher float charge causes higher heating and gassing of the battery. Higher gassing results in more release of hydrogen. The excess release of hydrogen in turn causes water loss. In addition, the positive grid corrosion rate is accelerated and leads to the temperature derating of the battery. For example, 50% reduction in life for each 15°F increase in the battery temperature above 77°F. Loss of water due to high-temperature operation also accelerates the dryout of battery cells. This condition shortens the battery life even further for both AGM and gel based VRLA products.

The developments in charger technology include temperature compensation, fold back protection to reduce the effects of higher temperatures, excessive gassing, and thermal runaway. Provisions for airflow between cells allows for more uniform thermal distribution from cell to cell thus maintaining a uniform temperature gradient in the battery pack.

Repeated cycling of a battery is essential for a long life. Users often mistake the expected float warranty to be the actual battery life under cycling conditions. In either case, the VRLA batteries can provide excellent cycle life along with a cycle warranty. Users should always request that a 20-year VRLA battery make 1,200 cycles to 80% depth of discharge at C/8 rate to 1.75.

In several battery designs, antimony serves to provide a higher energy density.

However, calcium on the other hand when used in float applications batteries tends to improve the battery life. This is owing to the slower positive grid corrosion and battery requires less maintenance. The addition of tin (Sn) to the calcium (Ca) alloys in VRLA batteries results in excellent cycling capability.

Antimony additives in the VRLA battery are similar to the flooded cells. The antimony additives in the VRLA battery cause the float current and gassing to increase as the battery undergoes cycling. At the end of its life, an antimony battery's float current is almost six times greater than that of a battery with calcium as an additive. Thus calcium-based batteries are suited for telecom based applications since they do not dry out as easily and exhibit lower positive grid corrosion.

EFFECTS OF EXCESSIVE HEAT ON BATTERY CYCLE LIFE

Depending upon the type of the battery, excessive heat generated during the charging and discharging process has a detrimental effect on the formation of the battery. In the case of VRLA batteries, the deterioration leads to the separator and electrolyte breakdown resulting in the gassing of the electrolyte. During the gassing process, the vents let the electrolyte evaporate and the cell dries out. This effect is accelerated at higher operating temperatures.

Applying a general rule of thumb under high charge rates, the life of a traction battery is reduced by half for every increase in the cell temperature. Thus the temperature-time integral affects battery life with the worst case being subject to high temperatures for an extended period of time. The effects of excessive battery charging, as illustrated in Figure 2–4, lead to void formations, significant gassing of the electrolyte, and electrode overheating, which in turn leads to oxidation of the cathode.

Similar to detrimental factors associated with excessive battery charging, inadequate battery charging leads to sulfation (sulfate formation) at the electrodes, which in turn leads to reduction in battery charge capacity as illustrated in Figure 2–5. The reduction in the charge capacity in turn leads to gassing of the electrolyte. This is due to the inability of conversion of the charge into useful energy at the electrode.

BATTERY STORAGE

All traction batteries undergo a gradual self-discharge over a period of time either during use or in storage. The loss of useful battery capacity is typically due to parasitic reactions that occur within a rechargeable battery cell. This loss rate (self-discharge) is a function of the cell chem-

Figure 2–4   Failure modes associated with excessive battery charging.

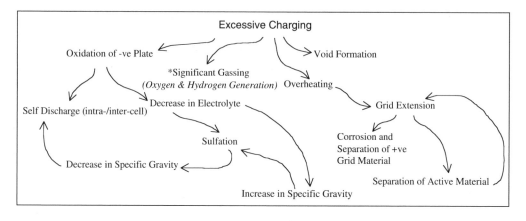

Figure 2–5   Failure modes associated with inadequate battery charging.

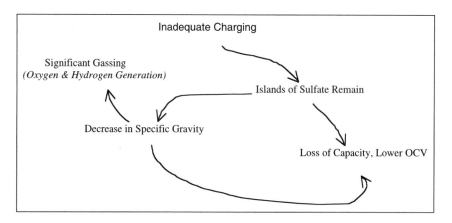

istry and the temperature environment experienced by the cell. Owing to temperature sensitivity of the self-discharge reactions, relatively small differences in storage temperature result in large differences in self-discharge rate. Thus, a traction battery pack when left in storage, with load connected, not only hastens the rate of self-discharge, but may also cause chemical changes after the cell is discharged, leading to cell reversal.

Storage of a VRLA Battery

As illustrated in the Figure 2–6, when a fully formed VRLA battery is kept in storage, the open circuit voltage (OCV) drops due to self-

Figure 2–6 Failure modes associated with battery storage conditions.

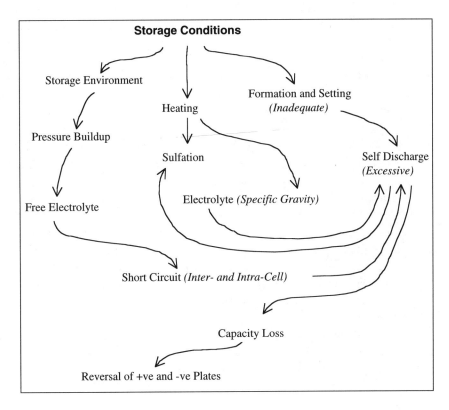

discharge. The drop in the OCV due to self-discharge in turn leads to battery failure over a storage period. In order to prevent failure of the battery in storage, it is essential to periodically apply a conditioning charge to the battery.

Storage of an NiMH Battery

The common rule of thumb for NiMH batteries is that an increase in storage temperature halves the time required to self-discharge to a given depth.

Traction batteries will provide full capacity on their first discharge after removal from storage and charging with standard methods. Batteries stored either for an extended period of time or at elevated temperatures may require more than one charge/discharge cycle to attain prestorage capacities. It is advised that special storage procedures should be obtained from the manufacturer in case the batteries are going to be in prolonged storage or rapid restoration of full capacity is desired.

NiMH batteries during the course of their storage, requires a recharge to maintain the OCV. For an NiMH battery, the charge factor is 1.5 to 1.6 and is expressed as ratio of the current capacity (Ahr) of NiMH battery to the capacity (Ahr) of the NiMH battery during the previous discharge. Since the gas recombination occurs on the activated surface of the NiMH battery, the interior of the NiMH battery cell electrode is not exposed to oxygen recombination. The high rate discharge is attributed to the minimal surface-only exposure of the NiMH battery electrode surface.

In addition, the self-discharge of a NiMH battery is significant for a battery stored at 100% SOC under room temperature conditions as shown in Figure 2–7. When NiMH batteries are stored under load conditions, small quantities of the electrolyte can seep through the seals or also through the vents. This creep leakage may result in formation of potassium carbonate crystals, which are obtrusive in appearance. In some cases, the creep leakage of an NiMH cell can result in corrosion of adjacent cells. Thus, it is advisable to isolate the battery electrically using an insulation tape on the positive terminal or silicone grease.

The key determinant of the EV battery system efficiency is the cycle life. As shown in Figure 2–8, cycle life is the acceptable capacity of the NiMH battery system. The limiting mechanisms are a combination of abrupt failure events as shown in the failure modes, typically related to charge, discharge characteristics. On the other hand, failures due to

Figure 2–7   Self discharge of NiMH battery stored at 100% SOC.

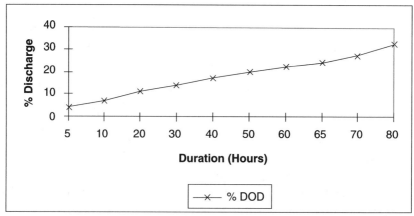

Figure 2–8 Variation of NiMH battery capacity and pressure with cycle life.

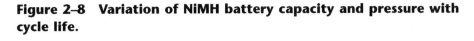

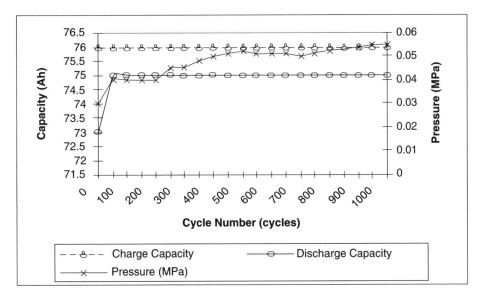

mechanical events resulting in the cell shorts or open circuits are relatively rare and random.

NiMH cell deterioration typically occurs owing to:

- Increase in the oxidation of the negative active material and increase in the internal resistance of the cell. This increase results in the reduction of the available voltage from the cell. This reduction is termed as the mid-point voltage (MPV) depression and affects the chemical equilibrium or balance between the cell electrodes. A reduction in the useful voltage possibly reduces the gas recombination capability, increases the intra-cell pressure, which ultimately leads to venting of the battery.
- A deterioration of the positive electrode active material results in the diminished chemical reaction and a consequent loss of useful battery capacity. The failure modes of the NiMH battery associated with loss of battery hydrogen storage capacity are shown in Figure 2–9.

THE LITHIUM-ION BATTERY

The charge/discharge cycle lives of the Lithium (Li)-ion battery cells containing lithium anodes tend to be terminated prematurely by short-

Figure 2–9   Failure modes of the NiMH battery associated with loss of hydrogen storage capacity.

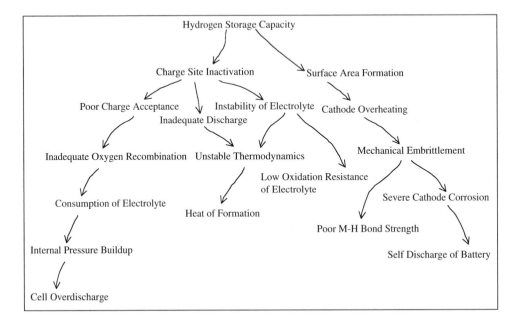

circuiting caused by the growth and penetration of lithium dendrites through the anode/cathode separators.

The approach to reduce such failures due to a short-circuit is by using lithium alloys and related lithium-intercalation materials. The related lithium-intercalation materials exhibit lithium activities lower than those of pure lithium.

Carbon-based materials are suitable for anode lithium-intercalation. This is owing to their low equivalent weights, low voltages versus lithium, and higher chemical stability in the presence of electrolytes. However, the carbon based materials, for example, graphite (has two times the lithium capacity as coke), are vulnerable to short-circuiting by dendrite growth. This occurs during the charging process because its potential is too close to that of the element—lithium. Additionally, developmental problems are further complicated by the fact that the electrochemical properties of carbon-based materials depend strongly on which particular electrolyte is being used. Mg_2Si is selected as a lithium-intercalate anode material because its electrochemical behavior is similar to that of carbon-based material. In addition, the intercalation of Li in Mg_2Si is highly reversible.

TRACTION BATTERY PACK DESIGN

Before commencing battery pack design for EV applications, it is important to understand the behavior of the battery chemistry. It is important to start with the requirements to ensure that the battery design falls within the boundaries of the technology. It is important to determine the maximum discharge rate of the battery in addition to the operating and storage temperature ranges.

Consider the cell configuration—Pb-acid, NiMH, and Li-ion batteries can be configured in the series or parallel configuration or a combination of the series-parallel combination. The selection of the configuration is based on the requirements for input voltage and battery pack discharge requirements. Battery cells can be configured in series to provide the necessary input voltages. Alternately, battery cells can be configured in parallel to provide the battery capacity to yield the required run time. Battery configuration can be determined based on the two methods.

Based on the lower battery cutoff voltage (LCV) and the minimum voltage from the battery (e.g., 12.5 V), the number of cells connected in a series combination is expressed by the equation,

$$LCV/12.5\,V = \text{\# of cells in the series combination}$$

Based on the average discharge current (I_{avg}), the number of batteries connected in parallel are expressed by the equation,

$$(\text{Run Time [hr]} \times I_{avg})/\text{Battery Ahr} = \text{\# of batteries in a parallel combination}$$

In the second method, the average discharge power (P_{avg}) is known and is determined by the equation,

$$(\text{Run Time} \times P_{avg})/\text{\# of series batteries} \times \text{Battery W-hr} = \text{\# of parallel batteries}$$

Battery pack electronics includes pack control (and possibly battery charge control). In addition, a simple fuel gauge or smart-battery circuitry is integrated into the battery pack. Contacts for the battery terminals must be designed to eliminate shorts by coins or other metallic objects. In addition, the contacts must exhibit good corrosion resistance and provide low electrical resistance.

Multilevel battery pack-control circuitry ensures reliability and provides protection against overcharge, over-discharge, short circuits, and thermal abuse. Charge-control electronics is typically part of the battery pack and is located outside the battery pack.

Estimation of the battery pack size may be performed during the design phase based on the electrode size (15 cm × 30 cm). In addition, battery pack weight may also be estimated for the EV. The battery electrodes are capable of yielding 30 Ahr.

Submodule design dimensions: 15 cm × 30 cm × 11.5 cm
Electrode size: 14.7 cm × 29.7 cm = 67.7 in²
Electrode capacity: 500 mAhr/67.7 in² = 30 Ahr
Volumetric energy density: 50 V × 30 Ahr/15 kg = 100 Whr/kg
Gravimetric energy density: 50 V × 30 Ahr/(1.15 dm × 1.5 dm × 3.0 dm)
                                     = 1500 Whr/5.175 dm³ = 290 Whr/L

The weight components for the 1.2 V unit cell battery pack are:

Cell Components	Thickness (mm)	Weight (g)
Conductive Plastic	0.10	4
Negative Electrode	0.85	85
Separator	0.25	12
Positive Electrode	1.08	117
Conductive Plastic	0.10	4
Total	2.38	222

For a 27 kWhr (300 V × 90 Ahr), the submodules are rated at 50 V, 30 Ahr. Overall nominal battery dimensions are 15 cm × 120 cm × 69 cm, which allows for a battery pack configuration with a low-profile battery allowing a low center of gravity.

The sub-module capacity is 30 Ahr × 50 V = 1,500 Whr
Dimensions: 15 cm × 30 cm × 11.5 cm
Weight: 15 kg
The module capacity is 30 Ahr × 300 V = 9 kWhr
Dimensions: 15 cm × 30 cm × 69 cm
Weight: 90 kg

The battery module design: 90 Ahr × 300 V = 27 kWhr
Dimensions: 15 cm × 120 cm × 69 cm = 125 litres
Weight: 360 kg (792 lbs)
Battery volumetric energy density is 27 kWhr/125 ltr = 216 Whr/ltr
Battery gravimetric energy density is 27 kWhr/360 kg = 75 Whr/kg

3 ELECTRIC VEHICLE BATTERY CAPACITY

The valve regulated lead acid-battery (VRLA) is a maintenance-free lead acid battery operating on the principle known as "sealed, recombination," wherein all the electrolyte is stored in absorptive glass mats (AGM) separators. The battery must remain sealed for its entire operating life and, to achieve maximum cycle life, must be properly recharged to prevent any excessive overcharge. Excessive overcharge results in excessive gas pressure build-up inside the battery, which is relieved by the opening of the pressure relief valve (typically set at 1.5 psi ± 0.5 psi). Everytime the valve opens, water vapor is lost, which in turn reduces battery life.

The battery has been developed from extruded lead onto glass-fiber filaments that are woven into grids (mats) for use as electrode plates. This process provides the desirable crystal structure of lead oxide (PbO_2) active material. The battery must be maintained, however, under optimal driving conditions.

The USABC has outlined the performance requirements for VRLA batteries for the near term and the next few years, especially for use in electric vehicle applications. These requirements are summarized in Table 3–1, which shows that the near term VRLA battery provides up to 95 Whr/L of energy, while the requirements are to increase the energy density to 135 Whr/L over the next few years. This increase in the energy density means that there has to be a significant increase in the battery capacity.

BATTERY CAPACITY

The useful available capacity of the battery (in Ahr) is dependent on the discharge current. This relationship can be expressed in the form

$$I^n \times t = K$$

where I is the discharge current in A, t ($0.1 < t < 3$) is the duration of the discharge in hours and n and K are constants for a particular battery type.

43

Table 3–1   Current and future VRLA battery specifications.

	Current Pb-acid	Midterm Specifications
Specific Energy	45 Wh/kg	80 Wh/kg
Energy Density	95 Wh/L	135 Wh/L
Specific Power	245 W/kg	150–200 Wh/L
Recharge Time	<10 hrs.	<6 hrs.
Cycle Life (C/3)	600	600

Figure 3–1   The estimated Peukert plot at 80°F for an 80 Ahr battery.

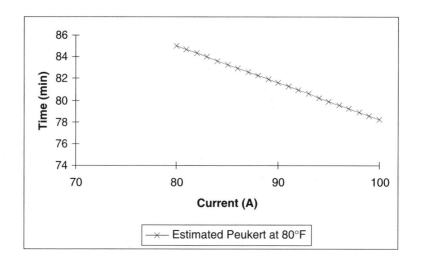

For example, an 80 Ahr VRLA battery, Peukert constants n may vary between 1.123 to 1.33 and K may vary between 138 to 300 respectively. The graph in Figure 3–1 is a Peukert plot at room temperature, 80°F.

THE TEMPERATURE DEPENDENCE OF BATTERY CAPACITY

The useful Ahr capacity available from the VRLA is dependent on battery temperature and may be represented by the following equation

$$C_t = C_{77} \times (1 - 0.065(77 - t))$$

where t is the temperature in °F, C_t is the battery capacity at t °F and C_{77} is the capacity of the battery at 77°F (room temperature). For example, C3 capacity at 32°F, for a 80 Ahr VRLA battery is expressed as C_3 (32°F).

$$C_3(32°F) = 80 \times (1 - 0.0065(77 - 32)) = 56.6 \, Ah$$

Similarly $C_{0.1}$ at 80°F for a 80 Ahr VRLA battery is 36.3 Ahr. Thus $C_{0.1}$ at 120°F is 45.74 Ahr.

Thus under a constant current discharge and variation of temperature the battery pack capacity changes the performance of the electric vehicle (EV). This is observed as a variation of the driving distance before an EV recharge.

As illustrated in Figure 3–2, the graph is the estimated VRLA battery capacity with respect to the battery pack temperature. A 80 Ahr VRLA battery above room temperature, 77°F, exhibits a larger than rated battery capacity. (See Figure 3–3.) This increase is larger at higher temperatures. A fully charged battery pack when discharged at 100°F can deliver approximately two times the rated battery pack current than a battery pack at room temperature under similar discharge conditions.

Figure 3–2   Variation of estimated VRLA battery capacity with temperature.

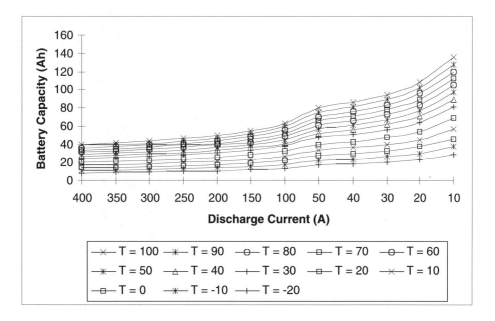

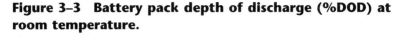

Figure 3–3   Battery pack depth of discharge (%DOD) at room temperature.

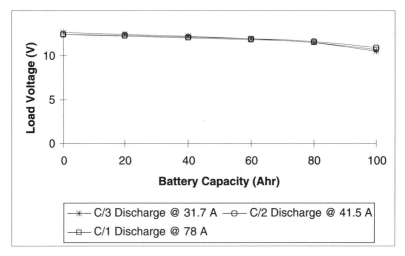

However, if this discharge is not regulated, it leads to premature battery failures due to deep discharge, leading to loss of electrolyte and gassing of the VRLA battery.

As the battery pack discharge rates vary, so does the battery pack performance. An identical depth of discharge (%DOD) may be achieved, at room temperature, using different rates of discharge at varying current levels. A fully balanced battery pack consisting of 80 Ahr VRLA batteries, delivers identical performance at C/3, C/2, and C/1 discharge rates.

STATE OF CHARGE OF A VRLA BATTERY

The state of charge (SOC) of a sealed VRLA battery is defined as the percentage of full charge capacity remaining in the battery. This information is identical to the combustion engine fuel gauge. In case of the EV, the SOC provides an indication of the amount of electrical energy remaining in the battery pack. The SOC is accurately determined by the measurement of the stabilized open circuit voltage (OCV).

Unfortunately, the VRLA batteries require at least five hours to stabilize after the recharge is applied.

Typically, the 100% SOC OCV for an 80 Ahr battery ranges between 12.9 V to 13.0 V (under room temperature conditions). The 0% SOC OCV for an 80 Ahr battery is 11.9 V. The approximate linear relationship between OCV and SOC may be expressed by

$$SOC = 84 \times OCV - 984$$

where 11.9 V < OCV < 13.0 V. The minimum allowable SOC is 20% under room temperature conditions. Determination of SOC in dynamic driving conditions is difficult owing to the additional OCV that influences the battery condition. The SOC under dynamic conditions can be expressed as

$$SOC = f_1 (V_{OCV}) + f_2 (I \times f_1 (V_{OCV})) + f_3 (\Delta T)$$

where f_1, f_2, and f_3 are functions of OCV, discharge current (I) and temperature ΔT. A good SOC calculator provides the following advantages for EVs:

- Longer battery life
- Better battery performance
- Improved power system reliability
- Avoid no-start conditions
- Reduced electrical requirements
- Smaller/lighter batteries
- Improved fuel economy
- Prefailure warning of the battery pack
- Decreased warranty costs

The SOC calculator monitors battery pack voltage, current, and temperature. The SOC can provide useful information about the absolute SOC, relative SOC, capacity of the battery pack, and the battery low acid level (if applicable).

The SOC calculation can be displayed in terms of useful battery capacity, health of the battery pack. The calculations provide optimization of recharge without detriment to battery life. SOC prevents overcharging of the battery when fully charged and prevents accidental discharge. As a battery replacement indicator, the SOC can warn the user that the battery pack capacity is at its threshold and requires charging. As a restart predictor, the SOC predicts at which temperature the battery pack may not be capable of cranking adequately. The SOC controls cooling fans and heater in the battery pack.

SOC calculations benefits include:

- Low SOC increase recharge rates by 25%—hot weather battery tests show up to 30% increased recovery rates and cold weather simulations demonstrate an initial charge rate increase of 40%.

• Maintaining a 5 to 10% higher battery charge level under severe conditions improves the useful battery capacity and life proportionately. Cold weather simulations demonstrate a 5% higher SOC with a 0.45 V charge increase. EVs with a 20% lower operating SOC are correlated with up to 50% shorter battery pack life.

• Reduced battery pack charge voltage at high SOC reduces the gassing and corrosion of the battery electrodes by approximately 40%. A 10% water loss and corrosion reduction is achieved per 0.1 V charge voltage. This reduction of water loss and electrode corrosion assumes that a high battery pack charge level is maintained for 80% of the duration of the charge and a low charge level is maintained for up to 5% of the duration of the charge.

Average battery pack life is projected to increase by approximately 30%. This improvement is attributed to the 40% reduction of battery pack water loss and electrode corrosion. In addition, an improvement in the battery pack life is attributed to a 5% higher SOC. Figure 3–4 compares the normal battery pack SOC versus the regulation battery pack SOC with respect to time. In addition, the battery pack capacity varies with regulation. The battery pack discharged to 50% SOC using SOC regulation provides an improved battery performance in comparison with normal regulation as shown in Figure 3–5.

Figure 3–4   Comparison of normal SOC and regulated SOC.

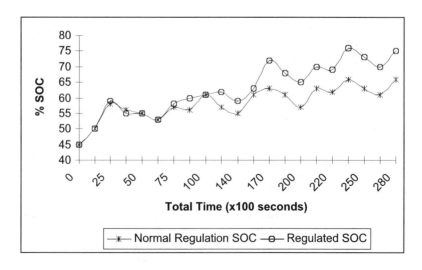

Figure 3–5   Comparison of battery pack capacity at 50% SOC.

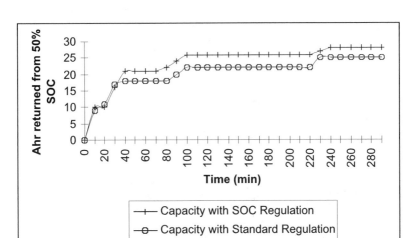

Practical State-of-Charge Calculation

The battery SOC can be estimated at each time interval in an iterative manner.

Estimate the battery voltage ratio by interpolating the battery SOC with respect to V_{oc} (open circuit voltage). Find the voltage ratio corresponding to the current SOC and multiplying the result by the rated voltage. Calculate the average voltage for the time interval by the average voltage (V_{ave}). The average battery voltage V_{ave} is expressed as

$$V_{ave} = \tfrac{1}{2}(V_0 + V_1)$$

where V_0 is the voltage at the beginning of the time interval for the SOC measurement.

Estimate the internal resistance R_{int} of the battery by interpolating within the SOC with respect to the battery resistance table. Calculate the average battery resistance (R_{ave}) for the time interval. The average battery resistance R_{ave} is expressed as

$$R_{ave} = \tfrac{1}{2}(R_0 + R_1)$$

R_0 is the battery resistance at the beginning of the SOC measurement interval.

Calculate the battery current I using the following equation

$$r = [V_{avg}/(2R_{avg})]^2 - P_{batt}/R_{avg}$$

If r is estimated to be greater than zero, then the current I, is estimated using the following equation.

$$I = V_{avg}/(2R_{avg}) - \sqrt{r}$$

Adjust the battery voltage using the equation $V = V_{avg} - (I \times R_{avg})$
Estimate the new SOC using the equation

$$SOC = SOC_0 - P\Delta t/3600 \times C \times V$$

Where P is the power derived from the battery and C is the battery's Ahr capacity. The number 3,600 appears in the divisor because the time interval Δt is expressed in seconds.

Repeat the calculation steps 1 through 6, as above, until the difference between the SOC_0 and the newly calculated SOC converges within 0.01% of the SOC.

Maximum Discharge Power

The maximum current drawn during discharge must not typically exceed 500 A. In a 100% SOC condition, the corresponding VRLA voltage will be approximately 11 V. Thus a 12 V VRLA battery can provide approximately 5.5 kW. In the case a 30-battery series string is connected together, the maximum power available to the powertrain would be 165 kW.

Maximum Recharge Power

The maximum current applied to a 12 V VRLA battery previously discharged to 20% SOC under room temperature conditions must not exceed 100 A. Correspondingly, the voltage of the VRLA battery should not exceed 15.5 V to prevent excessive loss of water vapor and irreversible damage to the battery. Thus, the maximum power applied to a single 12 V VRLA battery during recharge is 1.55 kW and the maximum power applied to a series string of 30, 12 V VRLA batteries is 46.5 kW.

Battery Output Power

During a constant current discharge at a C/3 current discharge (i.e., at a three-hour rate), the voltage profile can be estimated by the following equation

$$(V_{OC} - V_T)(A - t) = B$$

where V_{OC} is the open circuit voltage of the battery, V_T is the on-load voltage of the battery at time t (hours), A and B are the constants to be determined. This is a hyperbolic equation and the curve is identical to the voltage of a battery during a constant current discharge.

The values of A and B may be determined iteratively to provide a close approximation to the actual voltages during the rates of discharge between the six-minute rate and the three-hour rate as possible. The resultant equation is where V_t is the voltage after time t hours into the constant current discharge and T is the rate of discharge in hours. As an example for an 80 Ahr, the three-hour rate discharge would have the equation

$$(13.054 - V_t)(3.7 - t) = 2$$

for the VRLA battery $V_{OC} = 13.054\,V$ and A = current discharge rate + 0.7, B = 2.

This equation, along with the Peukert equation, provides a voltage through a discharge at varying currents taking into account the SOC. For the case of regeneration, it can be assumed that the battery is 95% efficient in accepting the regenerated charge current.

Each VRLA quickly develops its own personality during its formation cycling to the extent that each battery behaves differently during recharge. Thus it is necessary to provide an equalizing charge during the recharging process of a series string of batteries.

CAPACITY DISCHARGE TESTING OF VRLA BATTERIES

As recommended by most manufacturers and also industry standards (IEEE 450), a VRLA battery should be replaced if it fails to deliver 80% of its rated capacity. There is a very simple reason for the 80% of the rated capacity value. Based on the typical battery life curve of a lead-acid battery, once the battery capacity begins to deteriorate, the fall-off occurs at a rapid rate. A fully balanced, new traction battery pack will exhibit up to 95% of its rated capacity upon delivery because the active material on the battery plates are still undergoing formation. Once the active materials on the plates reach full formation, the battery capacity rises to its 100% capacity rating. This occurs and is maintained if the battery is under a proper state of charge, typically for a period of six months to several years. Capacity will continue to rise and will exhibit a rating of 100% for almost the entire battery life. As the plates begin to deteriorate and lose active material due to corrosion, loss of mechanical strength occurs, or electrolyte dry out begins, and the battery loses its capacity fairly rapidly. The deterioration of the battery is not as rapid at the end of life. Rather the deterioration begins at close to 80% of the capacity rating and falls-off rapidly from that point.

If a new battery pack is stored at delivery without a maintenance charge for an extended period of time, it may lead to the development of sulfation of the battery electrodes. This phenomenon will contribute to the additional loss of useful battery capacity during the service life of the battery pack. It is strongly recommended to place the battery pack on a maintenance charge as soon as possible. Also, the manufacturer of the battery must be consulted in case the battery pack exhibits less than 90% of its useful capacity.

The only accurate test of the useful battery pack capacity is the capacity discharge test. The test measures the amount of power removed from a fully charged battery over a rated time period. A capacity discharge test is performed on the battery pack while maintaining a constant current (or constant power) discharge on a battery bank using a regulated resistive load. The cell voltage and the battery pack voltage are monitored during the period of the discharge. Both the voltage values decrease over the discharge period. The time taken to reach the cell lower cut-off voltage (determined by the manufacturer) is noted and used to determine the overall battery pack's capacity.

Before the capacity test is performed on the battery pack, proper load and monitoring equipment is installed along with resistive load units. The resistive load units must be capable of manual or automatic control. Resistive load units are typically a series/parallel configuration of resistor banks with forced cooling fans. The resistor configurations allow adjustments of the load currents in small increments through relays and switches. In addition, a main circuit breaker switch provides protection, while ampere- and volt-meters monitor the cell/battery pack current and voltages, respectively.

The load units are specified and selected based on the Ahr rating and voltage levels of the battery pack. Data published by the battery manufacturer in the form of curves or tables for specific model types and Ahr ratings are a useful reference to determine the load unit specifications. The individual cell performance data sheets provide both discharge times that can range from one minute to eight hours with various constant current (or constant power) loads leading to a specified final battery voltage. Selection of a load unit with a large ampacity (ampere capacity) allows for shorter duration discharge tests simulating city driving conditions with greater user flexibility to profile the test.

The ANSI/IEEE 450 standard recommends that a minimum of three sets of readings be taken. One reading may be done at the beginning of the test, one reading upon completion of the test, and then one reading at an interval sometime during the test. This interval could be during the midpoint of the test. These battery pack discharge capacity tests will quickly identify the weak cells and also battery cells that are approaching reversal (displaying a 1 V or lesser voltage). The batteries exhibiting weak cells can then be removed from the battery pack and replaced by new batteries. A balancing charge should immediately follow the replacement of the bad batteries to balance the battery pack.

BATTERY CAPACITY RECOVERY

The cycle life of VRLA battery is directly dependent on the depth of discharge (DOD). In addition, the rate of charge of the VRLA also influences the battery life.

Battery cycle life is defined as the number of cycles completed before the discharge capacity falls below 15 Ahr (15 Ahr is defined as the battery end-of-life). Upon completion of the discharge, a reconditioning charge is applied using the following steps under room temperature conditions.

- Discharge the battery pack at the specified rate to specified depth of discharge
- Charge each battery at approximately 2.5 V per cell with specified current limit for a specified time
- Rest at open-circuit for the specified time
- Repeat the steps until the discharge capacity declines below 15 Ahr at the cutoff voltage of approximately 1.5 V per cell

If a balanced battery pack is maintained at low DOD, the battery cycle life improves to approximately 4,000 cycles. This condition can seldom be maintained for an EV owing to city driving patterns.

Formation of the passivation layer causes the active material to become electrically insulated and/or isolated from the grid. This limits the capacity of the battery available for discharge. The nature of the passivation layer depends on the type of battery grid alloy and by the electrolyte additives. Research results indicate that presence of phosphoric acid in the electrolyte and tin in the grid alloy reduces the passivation (passive reaction) effects.

Under the presence of higher currents, the passivation layer is highly porous. The degree of porosity is determined by the distance between the $PbSO_4$ crystals. In addition, larger current densities result in smaller PbO_2 particles. At higher current levels, the formation of a more porous surface layer on the positive grid. During the process of anodic polarization of the metal electrode, an insoluble anodic layer is formed at the surface of the electrode. This layer may be polycrystalline or a homogeneous nonporous film. Even at high charge current levels, the passivation layer builds up to the point where discharge capacity can be severely limited.

The formation of the crystalline layer is determined by the changes in potential and resistance. When the entire electrode surface is covered by $PbSO_4$ crystals, the potential of the electrode increases rapidly and the resistance remains constant. The electrode is passivated with an increase in the battery potential. This increase in the battery potential does not affect the capacitance and resistance values. The $PbSO_4$ layer tends to undergo a conversion to PbO_2. Under open circuit conditions, the battery potential takes values lying between the equilibrium potentials of the $PbSO_4$ and the $PbO_2/PbSO_4$ electrodes. Thus the VRLA battery undergoes $PbSO_4$ passivation in two ways: by anodic polarization of the electrode and by self-passivation under open circuit conditions.

In order to achieve the maximum cycle life from the VRLA batteries, it is both required that the DOD be kept at low as possible and that the charge current limit is as high as possible. This ensures that the passivation of the battery electrodes is at a minimum.

DEFINITION OF NIMH BATTERY CAPACITY

NiMH batteries are rated with an abbreviation C, the capacity in Ahr. The C rating for the NiMH battery is obtained by thorough conditioning of the individual NiMH cells. This can be established by subjecting the cell to a constant-current discharge under room temperature. Since the cell capacity varies inversely with the discharge rate, capacity ratings depend on the discharge rate used during the discharge process.

For NiMH batteries, the rated capacity is normally determined at a discharge rate that fully depletes the cell voltage in five hours. For the purpose of electrical analysis of the battery cell, the Thevenin equivalent circuit is used. This circuit models the circuit as a series combination of the voltage source (E_0), a series resistance (R_h = the effective instantaneous resistance), and the parallel combination of a capacitor (C_p = the effective parallel capacitance) and the resistor (R_d = the effective delayed resistance).

Figure 3–6   Recovery of battery cell discharge voltage.

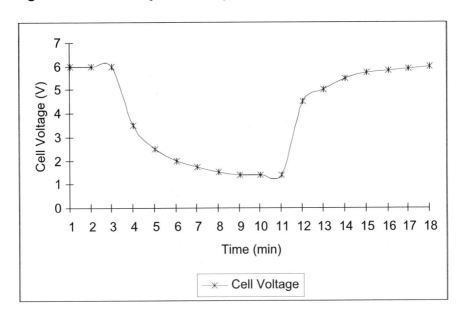

Under steady state conditions, the cell voltage at a known current draw is $E_0 - iR_e$, where R_e is the effective internal resistance of the NiMH cell. R_e is the sum of the R_h and R_d. Under transient discharge conditions, as shown in Figure 3–6, the initial voltage drops immediately to $E_0 - iR_e h$ and then rises exponentially, with time constant $Cp \times R_d$ to a steady state voltage. This discharge condition reverses once the load being applied is removed from the battery as seen in Figure 3–6 above. Note that the slow recovery of NiMH cell voltage after removal of the load after approximately 11 minutes is attributed to the delayed resistance R_d. This behavior is identical to the effect noticed during discharge between 4 and 11 minutes.*

For most applications, unlike EV applications, the steady state voltage is adequate for describing the battery performance. This is owing to the fact that the time constant for most cells is small—typically, the time constant is less than 3% of the discharge time. Although the instantaneous resistance of the NiMH cell is comparable with NiCd cell, the delayed resistance is approximately 10% higher. For this reason, the steady-state voltage for the NiMH cell is lower than that of NiCd.

*Note: This discussion is also made in Chapter 6 to describe the battery discharge characteristics.

NiMH Battery Voltage During Discharge

The discharge voltage profile for an NiMH cell is affected by transient effects, discharge temperature, and discharge rate. Under most conditions, the voltage curve retains the flat plateau before a rapid drop off termed as the knee of discharge curve, as observed between 80% and 100% discharge. A typical discharge profile for a cell discharged at a five-hour rate (0.2 C) results in the open circuit voltage drop from 1.25 V to 1.2 V. This discharge occurs rather rapidly. As seen by the flatness of the plateau and the symmetry of the curve, in Figure 3–7 the midpoint voltage (MPV—the voltage when 50% of the available cell capacity is depleted during discharge) provides a useful approximation to the average voltage available throughout the discharge cycle.

Figure 3–8 summarizes the operating voltages for a 90 Ahr NiMH battery. As the DOD of the NiMH battery varies with the discharge rate, the amount of useful current available from the battery and thus the battery pack decreases. The discharge of the battery pack is continued till the first battery in the pack is fully discharged and reaches the cut off voltage of 8 V.

Table 3–2 tabulates the operating voltage and voltage limits of the 90 Ahr NiMH battery with the operating voltage limit of 8 V.

Figure 3–7   Variation of midpoint voltage (MPV) with temperature.

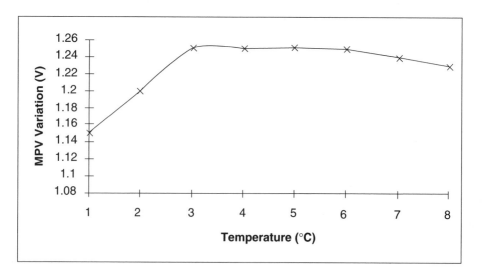

Figure 3–8   Variation of NiMH battery voltage with respect to the depth of discharge (%DOD).

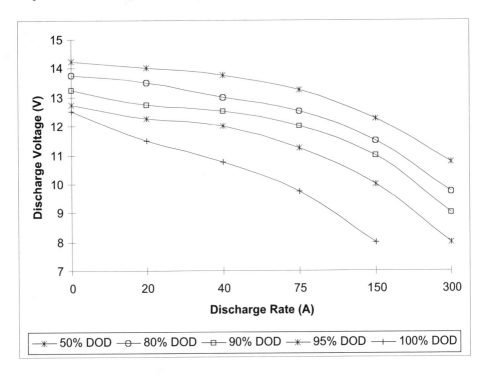

Table 3–2   Operating voltage and voltage limits of the 90 Ahr NiMH battery.

% SOC	% DOD	Discharged Capacity (Ahr)	Remaining Capacity (Ahr)
90	10	9	81
80	20	18	72
70	30	27	63
60	40	36	54
50	50	45	45
40	60	54	36
30	70	63	27
20	80	72	18
10	90	81	9
5	95	85.5	4.5
0	100	90	0

Effect of Temperature on Discharge

As noted earlier, the main environmental influences on the location and shape of the voltage profile are discharge temperature and rate of discharge. Small variations in the room temperature do not affect the NiMH cell voltage profile. However, large deviations, especially under lower temperatures, reduce the MPV of the cell while maintaining the general shape of the voltage profile. This results in a diminished useful capacity of the NiMH battery.

As seen in Figure 3–9, the battery pack resistance varies with DOD. Under city driving conditions, a 90 Ahr NiMH battery resistance drops from 13 mΩ to 12 mΩ as the DOD changes from 0 Ahr to 40 Ahr. Thus the battery pack is capable of delivering a higher discharge current at 40 Ahr under nominal operating temperature conditions.

LI-ION BATTERY CAPACITY

The theoretical specific capacity of the Li-ion active materials is 148 mAhr/g for $LiMn_2O_4$ and 372 mAhr/g for carbon. Thus at a mean

Figure 3–9   Effect of depth of discharge (DOD) on battery resistance.

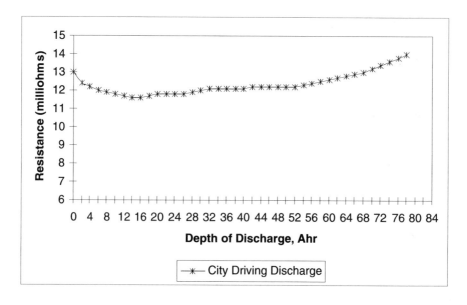

discharge voltage of 3.8 V the Li-ion battery provides a theoretical specific energy of 400 Whr/kg. The reversible capacity is reduced owing to the fact that all the lithium in the positive electrode is not used. In addition, during the first cycle, the lithium is irreversibly consumed due to passivation on the carbon side. Despite these reductions, the theoretical specific capacity is still around 300 Whr/kg.

In comparison with $LiNiO_2$ and $LiCoO_2$, $LiMn_2O_4$ is cheaper and more environmentally benign. In addition, the $LiMn_2O_4$ is more resistant to overcharge, since Mn(iv)—in contrast to Co(iv) and Ni(iv)—is stable. Thus making the $LiMn_2O_4$ cell intrinsically safe for use in the EV designs.

Although the conductivity rate of the electrolyte is two orders of magnitude lower than in alkaline systems, the full capacity of the Li-ion battery cannot be discharged at high discharge currents. Even at 1C discharge rates, more than 80% of the Li-ion battery can be discharged successfully. Continuous high currents are atypical in EV applications. In addition, short discharge pulses are required for acceleration of the EV. The Li-ion battery system is excellent with pulsed discharge applications—185 W/kg at 30 seconds of pulses down to 80% DOD of the prizmatic cells as shown in Figure 3–10.

Figure 3–10   Variation of peak Li-ion battery power with discharge.

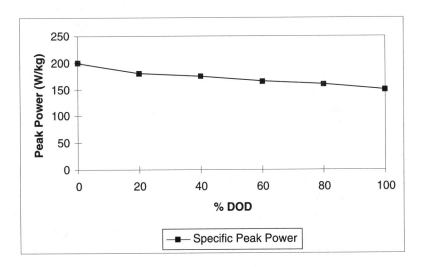

BATTERY CAPACITY TESTS

The capacity tests specified in the ANSI IEEE 450 standard are categorized into four tests,

- Battery pack acceptance test
- Battery pack performance test
- Battery pack service test
- Battery variable power test

The battery acceptance test is used to determine if the battery bank meets its purchase specification or the manufacturer's specification. This test is performed at the manufacturing facility or upon installation of the battery.

The battery performance test is performed periodically to measure the battery pack capability, including operation, age, deterioration, and environment. This test can be performed at any time during the entire life of the battery.

The battery service capacity test determines whether or not the battery system, as per manufacturer specifications, will meet the battery pack requirements during load and duty cycling. This test is not designed to effectively predict battery life or replacement, rather it is performed as part of the preoperational or periodic DC system check.

Before a battery pack undergoes an acceptance or performance test, a complete preventative maintenance inspection should be scheduled. The inspection of the battery pack should include measuring and recording the battery terminal resistance of all the connections in the pack. Resistance tests should be performed on the battery pack. The battery pack should be placed on a float charge for three days and not more than seven days prior to performing a capacity test to ensure that the battery pack is at a 100% SOC.

On the contrary, service tests are performed on the battery in an "as is" condition and do not require pretest conditions. For service capacity tests, the rated discharge current and the testing period should ideally match the duty cycle of the system. This means that the user should be prepared to manually or preprogram the test equipment to change the current discharge levels at specified time periods as required by the duty cycle.

For the acceptance and the performance capacity tests, a rated discharge current and the testing period are selected from the battery manufacturer's cell performance data sheet based on the battery model type, amp-hour rating, the load unit's current rating, and the final end voltage per cell. Typically, the voltage for a VRLA cell as selected to be 1.75 V per cell. This voltage is consistent with good engineering practice to take in to account DC device and inverter systems operating ranges. The ranges allow for a 10% variation of operating voltage levels.

The battery rated discharge current selected is then corrected to a test discharge current based on the average of the temperature readings that are recorded previously. The test discharge current is equal to the rated discharge current divided by a discharge correction factor.

The resistive load unit and battery monitoring equipment is set up and connected to the battery pack. For battery acceptance and performance capacity tests, a constant discharge current is maintained either automatically or manually when the load is switched on. The current drops to only 10 to 15% over the entire test period. The individual cell voltages and the battery pack's overall voltage are read and recorded at selected intervals during the test using data monitoring equipment. Time is kept, and the test is concluded when the terminal voltage decreases to a value equal to the final end voltage (typically 1.75 V) times the number of cells in the battery pack.

In case an individual cell approaches reversal of its polarity (defined by ANSI/IEEE 450 plus 1 V or less), and the final terminal voltage has not yet been reached, the test should be stopped long enough to remove the weak or reversed cell(s) from the string. When the weak or reversed cell(s) in the string are replaced, the discharge timing and the discharge test is continued. However, removal of the weak cell should be done as quickly as possible, within six minutes or 10% of the test time, whichever is shorter. This prevents the test results from being skewed by the weak or discharged cell. If the weak battery cannot be removed from the battery pack in time, the test must be aborted and rescheduled to allow the battery pack to be recharged and equalized. The possibility of the weak cells should be anticipated, and connectors with premade links should be made prior to when the discharge test is conducted. This eliminates delays that may result due to removal, replacement, and reconnection of the batteries in the pack. A new final terminal voltage should also be determined when continuing the test, based on the remaining number of batteries in the pack.

If a battery bank does not meet the requirements of the duty cycle, a complete maintenance inspection should be scheduled, and any necessary corrective actions should be taken. The battery pack's rating should also be reviewed based on the duty cycle of the load it is being applied to. Since a service test is performed on the battery in the "as found" condition to reflect the quality of the maintenance and operating practices, it cannot effectively predict battery life or battery replacement time.

The battery acceptance or performance capacity is determined by the equation

$$\% \text{ Capacity at } 25°C \ (77°F) = T_a/T_s \times 100$$

where T_a is the actual time to the final end voltage and T_s is the rated time to the final end voltage.

As suggested earlier a battery pack that demonstrates a pack capacity of less than 80%, should be replaced. As a normal design practice a VRLA battery is sized to include a 10 to 15% growth factor and a 25% aging factor. Oversizing a VRLA battery by more than an aggregate growth factor of 40% is not possible. This is due to the rapid deterioration of the battery end-of-life deterioration. In addition, the new battery pack delivery pack ranges between 6 to 14 weeks, depending on the battery size. Spare batteries to populate an entire replacement pack are generally not stored. This is owing to the self-discharge of the battery pack that occurs during long duration of storage.

The variable power discharge is a simplified version of the urban driving time power test. This power discharge effectively simulates the dynamic discharging and can be implemented in the laboratory as the simplified urban driving test. The 100% power discharge rating on the graph is intended to be 80% of the battery consortium's power goal.

The test may be performed based on the manufacturer's battery ratings. As an example, if this profile is scaled to 80% of the battery consortium goal of 150 W/kg the battery discharge will be at a peak power of 120 W/kg and an average power of 15 W/kg. A lower power version of this profile may be used for testing batteries that cannot be operated at the nominal peak power requirement.

The battery pack will be charged and the pack temperature is stabilized in accordance with the manufacturer's recommended procedures or per the battery test plan. The 360-second discharge test profiles are repeated end-to-end with no time delay (rest period) between them. The maximum permissible transition time between power steps is one second, and these transition times are included in the overall length of the discharge profile (i.e., a discharge test is always 360 seconds long). This discharge regime is continued until the end-of-discharge point specified in the test plan (normally the rated capacity in Ahr) has been reached. Alternately, up to the battery voltage limit, whichever occurs first, has been reached. If the maximum power step cannot be performed within the voltage limit and the specified end-of-discharge has not been reached, power for this step is limited to the weakest battery in the battery pack. This may be because the battery cannot sustain 5/8 of this power within the voltage limit (or the specified discharge is reached), at which point the discharge test will end. The end-of-discharge point is based on the net capacity removed (total Ahr–regeneration Ahr) from the battery pack. In this case, the test is terminated at the point that the first battery reaches the end-of-discharge point. The battery is fully recharged as soon as practical after the discharge. If the battery can be damaged as a result of regeneration, the power, current or voltage may be regulated during the discharge test. Table 3–3 shows a 20-step test profile, also known as the 360-second frame that simulates the driving conditions and is repeated until the weakest battery in the pack reaches 100% DOD.

Figure 3–11 illustrates the change in the resistance of an 85 Ahr VRLA battery with respect to the %DOD for a discharge test. The variation is the worst-case source resistance calculated for the performance of the battery pack when the EV goes from a cruising speed of approximately 30 mph (20 A) to a hard acceleration 60 mph (170 A).

Table 3–3 Twenty-step test profile, also known as the 360-second frame.

Step Number	Duration (secs)	Discharge Power (%)	Description (W/kg)
1	16	0	Rest
2	28	–12.5	–15
3	12	–25	–30
4	8	12.5	15
5	16	0	Rest
6	24	–12.5	–15
7	12	–25	–30
8	8	12.5	15
9	16	0	Rest
10	24	–12.5	–15
11	12	–25	–30
12	8	12.5	15
13	16	0	Rest
14	36	–12.5	–15
15	8	–100	–120
16	24	–62.5	–75
17	8	25	30
18	32	–25	–30
19	8	50	60
20	44	0	Rest

Figure 3–11   Change in resistance with discharge.

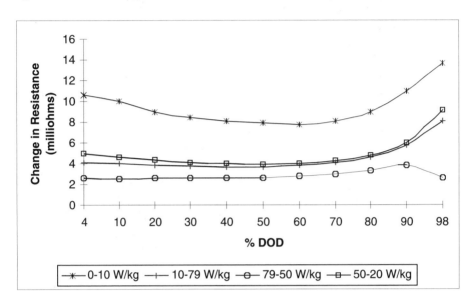

Identical to the VRLA battery, a 90-Ahr NiMH battery undergoes a similar change in the battery resistance. The discharge variation is the worst-case resistance calculated for the performance of the battery pack when the EV goes from a cruising speed of approximately 30 mph (20 A) to a hard acceleration of 60 mph (170 A). Figure 3–12 illustrates a change in the battery resistance for a regenerative charge applied during the driving at 15 W/kg and 60 W/kg, respectively.

ENERGY BALANCES FOR THE ELECTRIC VEHICLE

Several factors influence the EV energy balance. The energy removed from the traction batteries, defined as a positive power gain is expressed in kilowatts (kW). The energy consumption during the time interval is calculated as the total power loss multiplied by the time increment and is termed as a negative power gain expressed in kilowatts (kW).

The factors influencing the energy balance of the EV include:

- Aerodynamic drag losses
- Rolling resistance losses

Figure 3–12   Change in resistance with driving profiles.

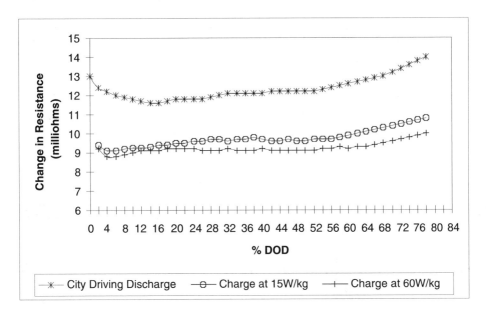

- Road inclination
- Power required for vehicle acceleration
- Transmission inefficiencies
- Power losses due to system controller (engine) inefficiencies
- Parasitic losses
- Power gained from regenerative braking
- Power from heat engine

The drag losses are associated with the EV body design. The power loss due to the aerodynamic drag, represented by a variable P_{aero} (watts) is expressed by the equation

$$P_{aero} = A_{frontal} \times C_{drag} \times V^3 \times \rho_{air}/2$$

where $A_{frontal}$ is the frontal vehicle area (m^2), C_{drag} is the drag coefficient of the EV, V is the velocity of the EV (m/s), and ρ_{air} is the atmospheric density (kg/m^3).

Rolling Resistance Losses

The rolling resistance is associated with the force necessary to overcome the friction of EV tires. The rolling energy loss equation required to overcome rolling resistance, expressed as P_{roll} is

$$P_{roll} = M_{Gr.Veh.} \times g \, (R_0 + R_1 \times V + R_2 \times V^2 + R_3V^3) \times V$$

where $M_{Gr.Veh.}$ is the gross vehicle mass (kg), g is the acceleration due to gravity (m/s²), and R_0, R_1, R_2, and R_3 are rolling resistance coefficients.

Road Inclination Losses

The following equation calculates the power loss associated with the road inclination. Expressed in watts, the road inclination loss (P_{incl}) is represented by the equation

$$P_{incl} = M_{Gr.Veh.} \times g \times V \times \sin \, (\beta_{incl} \times \pi/180)$$

where β_{incl} is the road inclination angle expressed in degrees with respect to the horizontal and converted to radians in the equation.

Vehicle Acceleration Power Losses

The following equation calculates the power requirements associated with the EV acceleration. Expressed in watts, the acceleration power loss P_{accel} is represented by the equation

$$P_{accel} = V_{ave} \times M_{Gr.Veh.} \times a$$

where V_{ave} is the average velocity (m/s) expressed as $V_{ave} = 1/2(V_2 + V_1)$ and a is the acceleration expressed as $\Delta V/\Delta t$ (m/s²).

Transmission Inefficiencies

The power loss associated with the transmission inefficiencies is estimated by dividing the power required to put the EV into motion. It is expressed as the ratio of the sum of the power losses due to aerodynamic drag, rolling resistance, road inclination, and acceleration by the EV transmission efficiency. The transmission efficiency is determined from the drive train efficiency data and the torque data. The torque converter speed output is expressed by the equation

$$\text{Torque converter speed} = V \times G/\pi \times d \text{ and}$$

$$\text{Torque converter torque } (\tau) = P_{move}/G \times \omega$$

where d is the EV tire diameter (m), G is the transmission gear ratio, and ω is the wheel rotation rate (RPM).

The torque (τ) and the converter speed are calculated using the above expressions. Next, the torque data input table (output torque as a function of the output speed) is interpolated to determine the speed ratio corresponding to the output speed—torque combination. The drive train efficiency is interpolated from the drive train efficiency table as a function of the speed ratio.

Power Losses Due to System Controller/Engine Inefficiency

The engine efficiency is defined as the ratio of the engine power to the energy consumed by the EV. The amount of traction battery energy consumed by the vehicle at any time is inversely proportional to the controller's efficiency. The engine efficiency model is currently interpolated or extrapolated using a table of the controller's efficiency with respect to percent rated controller power. The efficiency is plotted as a function of the rated controller power, based on the fuel economy data (Figure 3–13).

Figure 3–13   Engine efficiency with respect to % rated controller power.

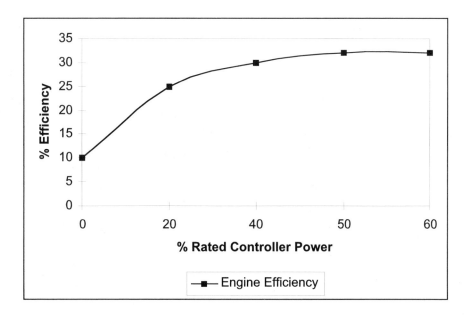

Power from Regenerative Braking

As the energy gained due to braking the wheels of an EV is returned back to the traction battery pack, there is a fractional gain of power. The regenerative braking gained is expressed as

$$P_{regen} = -e_{regen} \times M_{Gr.Veh.} \times a \times V$$

where a is the acceleration (m/s^2), e_{regen} is the regenerative braking efficiency.

Power from a System Controller/Engine

The energy consumed by the conventional combustion engine may be determined using the equation

$$Fuel = P_{engin} \times \Delta t/(H_{fuel} \times \rho_{fuel} \times e_{engine} \times e_{alternator})$$

where e_{engine} is the engine efficiency and $e_{alternator}$ is alternator efficiency.

In case of a conventional vehicle, the power from the engine (P_{engine}) is equal to the sum of all power losses. The loss P_{engine} is expressed by the equation

$$P_{engin} = P_{move} + P_{parasitic} + Fuel\ energy\ consumed$$

In case of an EV, the power from the engine may be determined using the equation

$$P_{engine} = P_{parasitic} + P_{move} + P_{regen} = P_{parasitic} + (P_{aero} + P_{incl} + P_{rolling})/e_{trans} + P_{regen}$$

where $P_{parasitic}$ is the parasitic losses, P_{move} is the total power to move the EV, P_{aero} is the aerodynamic drag loss, P_{incl} is the inclination loss, $P_{rolling}$ is the rolling resistance loss, and e_{trans} is the transmission efficiency.

In case of a hybrid vehicle, there are additional losses associated including the energy storage system and regenerative braking system. The energy is estimated as the integral of power and time. The energy losses and the energy gains are added to the SOC of the traction battery system. The engine power level is adjusted using the Auxiliary Power Unit (APU) control file.

4 ELECTRIC VEHICLE BATTERY CHARGING

Each VRLA quickly develops its own personality during its formation cycling to the extent that each battery behaves differently during recharge. Thus it is necessary to provide an additional charge during the recharging process of a series string of batteries. This charge is referred to as the equalizing charge.

CHARGING A SINGLE VRLA BATTERY

The constant current-constant voltage (CI-CV) algorithm is the suggested recharge method for the 12 V VRLA battery. It is recommended that this recharge be applied to the battery once every 20 to 30 battery charge-discharge cycles. The correct charging methods and optional values of current and voltage are presented in Figure 4–1.

The value of the charge current (I_c) depends on the output power delivered by the battery charger, the maximum and the minimum values of I_c, the corresponding values of the clamp voltage (V_c), and the time required to recharge the battery from 80% depth of discharge (DOD).

Each VRLA battery must remain sealed for life. In order to achieve maximum battery cycle life it is very important to prevent excessive overcharge. Overcharge results is gas pressure build-up, which is vented as water vapor. The loss of water vapor will result in drying of the cells and consequently reduce the life of the battery.

CHARGE COMPLETION OF A SINGLE VRLA BATTERY

The recharging of a VRLA battery using the CI-CV is completed when the current during the constant voltage phase has fallen to 0.80% of the battery's three-hour capacity (C/3). For example, a VRLA battery rated at 95 Ahr (rated for a C/3 rate) shall have attained a full

Figure 4–1   Multistep charging of a VRLA battery string.

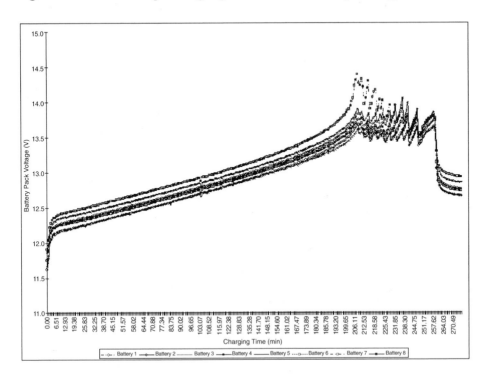

charge when the charge current has fallen to $0.008 \times 95 = 0.76\,A$ (approximately 1 A).

Temperature Compensation

Charging a VRLA battery increases the battery temperature above ambient. This temperature rise affects the cell detrimentally. The rise in temperature leads to gassing if the clamp voltage (V_c) attained during the charge is not reduced. For example, if the ambient temperature is above 77°F, V_c must be reduced by 0.01 V/°F above 77°F. If the ambient temperature is below 77°F, V_c must be increased by 0.01 V/°F below 77°F. The following example will illustrate the correct method of calculating V_c for a VRLA battery rated at 60 Ahr with a charge current of 60 A and an ambient temperature of 60°F.

The three-hour current delivering capacity (C/3) of a 60 Ahr battery is 60 A. I_c is 20 A, thus the ratio of I_c/C3 is 20/60 = 0.33. The corresponding V_c at 77°F for the ratio $I_c/(C/3) = 0.33$ is 14.53 V.

Since the ambient temperature is 77°F, V_c must be increased by $17 \times 0.012 V = .24 V$. The correct V_c for a 20 A inrush current for a 60 Ahr battery at 60°F is then $14.53 + 0.24 = 14.77 V$.

Overcharging of the VRLA Battery

A typical VRLA battery requires 3 to 5% of overcharge during the daily recharge using the CI-CV algorithm. If 50 Ahr are taken out from the battery at a 100% state of charge (SOC) condition, then approximately $50 \times 1.03 = 51.5$ Ahr must be returned to the battery during the next recharge to return the battery to a 100% SOC condition. As the battery ages, the degree of overcharge necessary to return the battery to a 100% SOC condition increases until, near the end of the battery's life, the battery will require 7 to 8% overcharge.

A good battery condition check is that the charge current at the point at which the battery has received 3% overcharge should be approximately 0.8% of the battery's three-hour capacity.

Equalization Charging of a Single VRLA Battery

As the VRLA batteries undergo cycling, their individual cells tend to fall out of step with respect to one another. For this reason, an extended constant current charge is required to balance the cells in the battery. It is recommended that an equalizing charge be applied once every 20 cycles. The values of I_c and V_c will be same during the equalization charge as the daily charge.

A battery reaches its 100% SOC when the voltage during the final I_{eq} phase does not rise more than 0.01 V during a 15-minute period. The duration of this equalization phase should not exceed six hours.

Recharging a Series String of VRLA Batteries

As VRLA batteries undergo formation cycles, their individual cells occasionally tend to fall out of step. This in turn results in an unbalanced pack of batteries when the batteries are connected in series. It is important to monitor the voltage and temperature of individual batteries in the series string. When the batteries reach a certain predetermined condition, the charge management system should send a control signal to change the charge current or modify the applied voltage based on the preset charging algorithm. In addition, the temperature compensation coefficient built into the charge management system should allow the charge current or clamp voltage to be regulated with any change of battery temperature.

Multistep Algorithm for Charging a Series String of VRLA Batteries

VRLA batteries, nominally rated at 12.0 V are best charged using a multi-step charging algorithm that follows these steps:

- **Step 1** A constant current is applied to the series string of VRLA batteries. The charging current is applied until the first battery (nominally rated for 12 V) in the string reaches a voltage of 15.5 V *or* until the last battery in the series string reaches a voltage of 14.5 V. At this point, the current being applied to the series string is reduced to approximately 50% of the initial start value to prevent loss of water due to gassing.
- **Step 2** The reduced constant current is applied until the first battery again reaches a voltage of 15.5 V *or* until the last battery in the series string reaches a voltage of 14.5 V. At this point, the current being applied to the series string should be reduced to approximately 50% of the current applied to start Step 2.
- **Step 3** The current is again reduced by half as in Steps 1 and 2 until the current being applied to the first battery in the battery pack is at 1% of the battery's three-hour rated capacity. For example, a 1% current of 90 Ahr battery is 0.90 A (approximately 1 A).
- **Final Step** The constant current is applied until all battery voltages have risen less than 0.01 V in a 15-minute time period. This equalization time period is important as it brings all the batteries in the pack within a 5 to 10% range of the first battery achieving the charge in Step 3. Furthermore, it is important that this step must not exceed six hours to prevent gassing of the batteries, resulting in the loss of water vapor.

TEMPERATURE COMPENSATION DURING BATTERY CHARGING

The on-charge voltage limits must be compensated for temperature to account for the variation of the useful battery capacity with tempera-ture. Table 4–1 summarizes the compensated lower and the upper

Table 4–1   Temperature compensated voltage limits.

Ambient Temperature (°F)	Lower Voltage Limit (V)	Upper Voltage Limit (V)
40	14.73	15.98
50	14.61	15.86
60	14.49	15.74
70	14.37	15.62
80	14.25	15.50
90	14.13	15.38
100	14.01	15.26
110	13.89	15.14
120	13.77	15.02

voltage limits with respect to the ambient temperature. Charging under ambient temperature over 120°F is not recommended.

Observations of the multistep algorithm for charging a series string of VRLA batteries:

Observation 1   Some of the batteries in the pack remain undercharged if the charging current is reduced when the first battery in the series string achieves a temperature compensated clamp voltage of 14.5 V.

Observation 2   Batteries equalize to within one cycle even with an unbalanced pack with improved battery capacity when an equalizing charge is applied to finish the charging for each individual battery. The charge is terminated when the first battery in the series string achieves a temperature compensated clamp voltage of 14.5 V.

Observation 3   Over gassing of the highest state of charge battery occurs when the battery pack is charged such that the last battery reaches a temperature compensated clamp voltage of 14.5 V but without an upper clamp of 15.5 V.

Observation 4   Battery equalization occurs before the equalization of the battery pack within three to four charge cycles with moderate gassing on some batteries. The battery pack is being charged using the last temperature compensated battery clamp voltage of 14.5 V with an upper clamp of 15.5 V. Higher gassing of high SOC batteries is avoided when lower capacity batteries are part of the series string. The 14.5 V voltage clamp results in less gassing than 15.5 V while maintaining good equalization of the battery pack.

CHARGING NIMH BATTERIES

At the beginning of the charging process, the NiMH cell is at room temperature, but as the charging progresses, the internal temperature rises up very rapidly owing to an exothermic reaction. During the reaction, the water in the electrolyte is decomposed into hydrogen atoms, which are absorbed into the cathode alloy.

$$\text{Alloy} + H_2O + e^{-1} \leftrightarrow \text{Alloy} \langle H \rangle + OH^{-1}$$

At the anode, the charge reaction is based on the oxidation of the nickel hydroxide.

$$Ni\langle OH \rangle_2 + OH^{-1} \leftrightarrow NiOOH + H_2O + e^{-1}$$

The exothermic reaction gives off heat throughout the charging process, and the cell temperature rises very rapidly. During this process it is easier to breakdown the electrolyte than it is to convert uncharged material into charged material.

RATE OF CHARGE EFFECT ON CHARGE ACCEPTANCE EFFICIENCY OF TRACTION BATTERY PACKS

The traction battery is charged until the pressure of the battery increases to 0.15 mPa at a constant current, ranging between 10 to 30 A. Before the battery reaches the overcharge range, the battery generates heat due to the exothermic reaction. Even in the range of overcharge with the charging current at 30 A under ambient temperature conditions, the pressure rise due to charging is appreciable. The charge acceptance of a string of NiMH traction batteries is improved as the charging rate is increased.

Temperature Sensing of Traction Battery Packs

Temperature of a traction battery can be measured using an NTC thermistor or other temperature sensing device that is attached to the surface of the cell. This method of temperature sensing relies on the transfer of heat due to conduction. The heat generated inside the cell is measured by the sensor attached to the surface of the cell. However, this measurement is often skewed by loss of heat from the cell surface due to convection and radiation, and depends upon the ability of the battery to transfer heat generated at the core of the cell to the surface. This factor

Figure 4–2 Variation of battery voltage and pressure during charging.

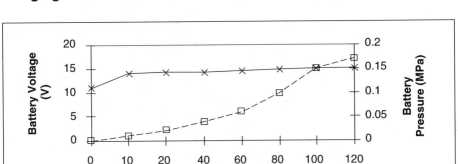

is also known as the thermal impedance of the battery varies with construction and cell size. The variation of the battery capacity under different battery profiles is illustrated in Figure 4–3.

When measured at the same charge rate, large size cells generate more heat than smaller size cells. This is because heat capacity of a smaller cell is higher than that of the larger cell. For example, a large 3,000 mAh NiMH cell with a volume of 22,500 mm^3 has a capacity of 7.5 mm^3/mAh in comparison to a 350 mAh NiMH cell with a volume of 3,150 mm^3 that has a capacity of 9.0 mm^3/mAh. This is an important design factor to consider when moving from battery packs with lower capacity cells to battery packs with higher capacity cells.

The temperature variation for small and large battery cells is illustrated in Figure 4–4.

The rate of change of battery temperature and the temperature difference are important design factors to consider when trying to locate the temperature sensor. The sensor should be placed on the largest available surface area. The thermistor as part of the resistor divider circuit has fairly low current flowing through the divider. The low current flow prevents thermistor self-heating and current drain into the power supply. The power supply is required to be stable across a wide range of ambient temperature range. This stability is essential to avoid a change in the voltage drop across the thermistor, which can be incorrectly noted as a temperature change.

Figure 4–3 Variation of battery capacity under different charge profiles.

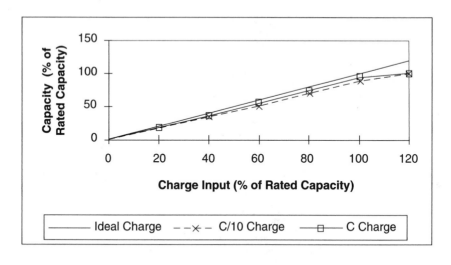

Figure 4–4 Temperature variation for small and large battery cells.

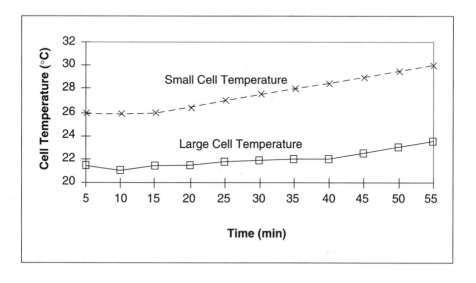

Thermistors have the advantage of being inexpensive, rugged, and sensitive to relatively small changes in temperature. In particular, negative temperature coefficient (NTC) thermistors are preferred, as resistance across the device decreases as the temperature being measured

Figure. 4–5 Battery pack thermistor voltage profiles.

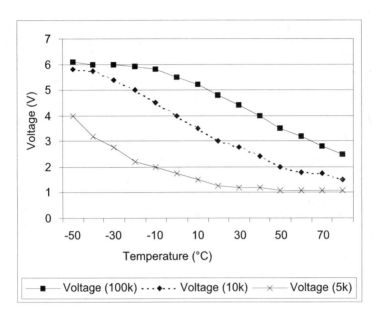

increases. This prevents an incorrect reading when the thermistor may be open exhibiting a high resistance to the divider circuit, which in turn suggests a large temperature change. Thermistor resistance specified at room temperature, temperature coefficient, and the thermal time constant are important selection parameters for battery sensor design.

Temperature ranges of thermistors for VRLA and NiMH batteries should run between −10°C to 25°C. Using a 5 V power supply, a 10 kΩ thermistor at 25°C is chosen. Although a choice of a higher-resistance value reduces the effects of lead resistance of the thermistor, it causes a shift in the effective linear temperature range. This shift must be compensated for by selecting a combination of thermistor and fixed resistance values that will allow it to operate in the relatively linear portion of the voltage-temperature characteristic. Figure 4–5 profiles three different thermistors between −50°C and 70°C.

Artificial battery heating and cooling mechanisms can affect the temperature sensor readings and thereby lead to over- or undercharging the battery. Since there is a large variation in the battery voltage and temperature as shown in Figure 4–6, the sensor used for monitoring battery characteristics must cover a wide range of battery voltage and temperature.

Figure 4–6   Variation of cell voltage and temperature with time.

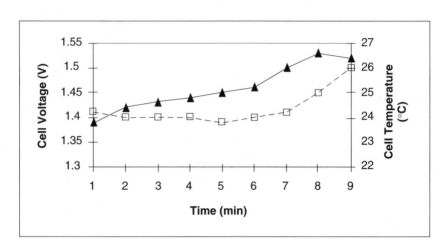

Temperature-Based Termination Methods

During the battery charging process, it is important to monitor the cell temperature to prevent cell failure. Some batteries during the process of charging may lack the drop in voltage—a commonly displayed phenomenon, which is usually used to detect full charge (refer to Figure 4–6). In case this drop is not noticed, the charge will not be discontinued. Therefore, a temperature-based termination method should be used as an alternate charge termination method. Three methods of charge termination are commonly used: maximum temperature cut-off, temperature change, and temperature slope (dT/dt).

The maximum temperature cut-off method is the simplest method of charge termination. Although cheap to implement, this method is not reliable. A simple circuit using a thermostat (bimetal thermal switch) or a positive thermal coefficient thermistor can terminate the charge current at the predetermined cut-off temperature. This mechanism assumes that the temperature is the true cut-off temperature—thus making an unreliable assumption that the batteries may be either hot or cold, resulting in overcharge or undercharge of the batteries. For NiMH batteries, the recommended maximum charge temperature is 50°C, as shown in Figure 4–7. In case the ambient temperature maintains the batteries under their maximum temperature and avoids the

Figure 4–7   NiMH battery cell voltage and cell temperature variation.

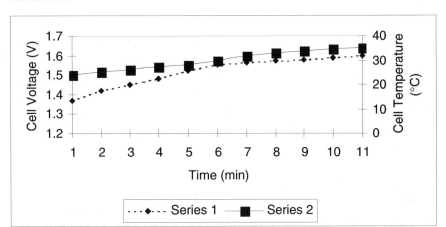

cut-off, an uncontrolled overcharge will result in permanent damage of the battery pack.

For these reasons, the maximum temperature cut-off charge method should not be used to determine fast charge termination. Instead this method should be a backup charge method to a more reliable primary charge method. This method can also be used as a temperature-sensing cut-off during long trickle charges and requires special care in selection of set points to avoid premature charge termination.

The second, temperature change method compares the difference between the ambient and the battery temperature. It requires two sensors monitoring the two temperatures, one sensor for the ambient temperature and the other for the battery temperature. This method may also prove unreliable in case the ambient temperature of the battery fluctuates. It may be necessary to compensate the fluctuations in the ambient temperature by attaching a thermal mass to the ambient temperature sensor. This limits the practical application of the temperature change method to applications where the ambient battery pack temperature is maintained at a relatively constant temperature. However, this method does provide an excellent back-up charge termination method.

The third most sophisticated charge switching method is based on the change of slope of the battery temperature profile as it eliminates the influence of the external environment. It can be a very effective technique for the early detection of overcharge. The cell temperature

Figure 4–8   Overcharge battery voltage and temperature characteristics.

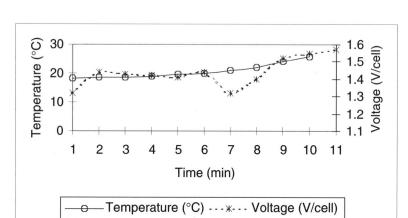

rises rapidly, indicating that an overcharge is occurring. If the method is sensitive enough, overcharge can be limited to a small temperature increase. An assumption made with the dT/dt method is that changes in the ambient will have a limited effect on the sensor relative to the heating of the cell due to overcharge.

The cell temperature, as shown in Figure 4–8, rises rapidly indicating that the battery is undergoing overcharge. In case the charging method employed is sensitive to a small change in the cell temperature, overcharge of the battery can be prevented. Although a change in the ambient temperature does not affect the slope dT/dt, cell heating due to overcharge causes a large temperature change in a short time period.

ENVIRONMENTAL INFLUENCES ON CHARGING

In the high temperature range between 40°C and 55°C, charging NiMH batteries require a careful selection of set points, for both temperature-based and voltage-sensing charging systems. Especially if the cell temperature has already been elevated prior to start of the charge. The charge efficiency of the NiMH battery drops at higher temperature owing to the release of oxygen at a lower SOC. The early release of oxygen is due to the decline of the voltage resulting in early release of oxygen.

During overcharge, the equation may be expressed as follows

$$OH^{-1} \rightarrow \langle 1/2 \rangle H_2O + \langle 1/4 \rangle O_2 + e^{-1}$$

Figure 4–9   Variation of battery capacity during charging acceptance inefficiency.

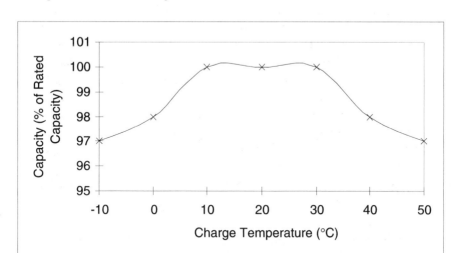

In the case of the sealed NiMH battery, oxygen gas evolved in the positive electrode during charge and overcharge process, is reduced to water by hydrogen in the negative electrode. The exothermic reaction causes a rise of battery temperature. The rise of battery temperature declines oxygen evolution voltage with a consequent increase in oxygen evolution. Owing to a polypropylene casing, the battery has a low radiation efficiency, which results in a low charge efficiency at elevated temperatures. The charge duration may also require to be extended due to charge acceptance inefficiencies as shown in Figure 4–9.

The charge time increases at lower temperatures so charge durations must be considered to provide adequate low-temperature charging, while avoiding excessive charge at normal temperatures. Charge rate must be reduced at low temperatures. An upper limit of 0.1C is recommended below 15°C and charging below 0°C is not advisable.

CHARGING METHODS FOR NIMH BATTERIES

Charging is the process of restoring a discharged battery to its original rated capacity. In order to be able to provide useful energy, the battery pack must be charged via the proper charge method. Various methods of charging are used to charge traction batteries.

The sealing principle of the NiMH battery suggests that the capacity of the hydrogen storage negative electrode is greater than that of the positive electrode. Thus the negative electrode is not fully charged when the positive electrode is fully charged. Thus any oxygen generated at the positive electrode during the charge and overcharge is chemically consumed at the negative electrode. This phenomenon suppresses the generation of battery internal pressure. However, during a rapid charge, internal pressure rises rapidly. The rise of the pressure actuates the safety vent and the electrolyte leaks. Loss of the electrolyte results in decrease in the electrolyte volume, lower discharge voltage, and lower cycle life.

The overvoltage (difference between the equilibrium voltage and the charge potential) tends to be greater when the charge current is higher. For a lower charge rate of 300 mA, the potential of the negative electrode reaches the hydrogen generation potential at the SOC of 100%. In contrast, as shown in Figure 4–10, at a higher charge rate of 1,000 mA, the potential of the negative electrode reaches the hydrogen generation potential at 50% SOC.

The first method utilizes an overnight charge (typically a slow charge) using a maximum constant current at C/3 A. This charging is a two-step process:

Figure 4–10   Variation of battery pressure during charging.

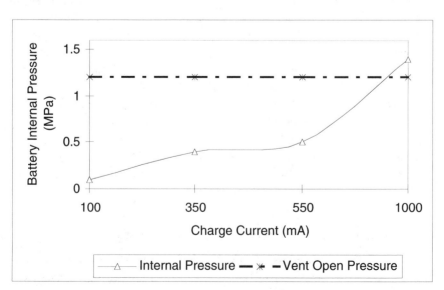

1. The first step is at a higher current, I_1 (the end of this first step is determined by the criterion, c1)
2. The second step is at a lower current, I_2 (the end of this second step is determined by the criterion, c2)

In the overnight charge, the first step current (I_1) and second step current (I_2) ranges between I_{1L}-I_{1H} and I_{2L}-I_{2H}, respectively. The allowed temperature rise is up to ranging between T_{1L}, T_{2L}, and T_{1H}, T_{2H} as the lower and upper temperature limits respectively. The first step charge is terminated once the temperature T_{1H} is reached, i.e., when the value of the slope dT/dt reaches the criterion c1(T) as shown in Figure 4–11 for a room temperature charge at the rate of 0.2 C.

The criterion c1(T) is a function of the actual battery temperature and is represented in Figure 4–12.

When the criterion C1 is reached, the first step of charge is stopped and the first step capacity charged is Ah_1. With zero rest period between steps, the second constant current I_2 charge step is initiated. The end of charge criterion C2 is reached when the capacity charged Ah_2 during the second step may be represented as

$$Ah_2 = mAh_1 + b \text{ where m and b are constant values.}$$

Figure 4–11   Slow charge temperature profile for NiMH batteries.

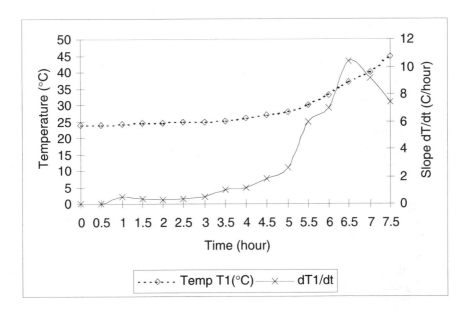

Figure 4–12   The c1(T) criterion for NiMH batteries.

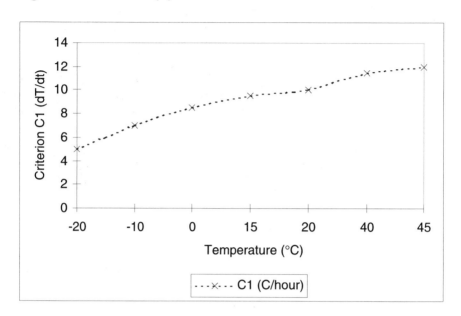

Using this charge mode, the end of the first step of charge is detected close to the theoretical end of charge at room temperature, 22°C.

The second charge method utilizes a fast charge one step constant current rate allowing the cells to recharge to 40% of their capacity from 20 to 40% initial state of charge. The battery packs are typically charged for 10 to 20 hours at over 0.5C A but less than 1C A and temperature ranging from room temperature to 55°C. As charging the batteries at a current in excess of 1C A causes internal cell pressure to increase resulting in the safety vent to be activated. This results in the electrolyte leakage. When the temperature of the batteries is under or over the commencement of the charge, rapid charge is terminated, and trickle charge is initiated.

Allowing high current to flow to excessively discharged or deep discharged batteries during the charge results in the formation of an electrode barrier that makes is difficult to restore the capacity of the traction batteries. It is important to first allow trickle current to flow, restore the battery voltage to its upper battery voltage limit control (1.8 V/cell approximately), and then proceed with the rapid charge current. This voltage is referred to as the rapid charge transition voltage restoration current and is normally 0.2 to 0.3C A.

Figure 4–13   Charge profile for temperature rise of 2°C/min.

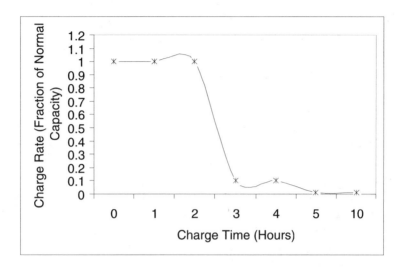

More precisely, when the battery voltage drops from its peak to 5 to 10 mV/cell during rapid charging, the charge is terminated and switched to trickle charge. In addition, the temperature of the traction batteries rises rapidly during the rapid charge. When the temperature rise of 2°C/min is detected, rapid charge is terminated and charge method is switched over to trickle charge as shown in Figure 4–13.

The overcharging of NiMH batteries, even with the trickle charge, causes a deterioration in the efficiency and cycle life. In order to prevent overcharging by trickle charging or any other charging method, the provision of a timer to regulate the total charging time is recommended.

Advances in NiMH Charging

In case the NiMH battery pack is charged too quickly, it can result in permanent damage to the battery and, in some cases, can also cause battery fires and explosions.

Combining efficient battery charging circuits and intelligent algorithms can improve the battery life and reduce the battery charging times. Battery designs must account for the variations in the vehicle design to allow for effective utilization of battery charging algorithms. The market for intelligent chargers is expected to increase by 20% by the end of 2001.

Fast charging of an individual battery or a battery pack refers to charging times in the one-hour range. Ultra-fast charging ranges from 5 to 15 minutes. It is important to charge the battery using an intelligent algorithm and extending the battery life, principally by not overcharging the battery.

A 5 or 15 minute battery charger would require in excess of 40C of charging capacity to charge a battery pack of 85 Ahr. (An 85 Ahr battery pack will require 85 A over a one-hour duration to fully charge.)

A high charge rate in excess of 40C will be extremely destructive on the battery chemistry. Temperature rise during charging will generate additional chemical reactions that are irreversible. Heat creates oxygen, which builds up pressure in a NiMH battery cell. This oxygen pressure leads to loss of water reducing battery life. Fast charging the NiMH battery monitors the cell's electrochemical condition during charging. The charging waveform is adjusted to produce the optimal charge acceptance. Instead of a preset waveform, the spacing of both positive and negative charging pulses is varied.

A more common approach for the NiMH battery charging is constant current charging. NiMH batteries exhibit an extremely flat voltage slope in comparison with the amount of charge that is in the cell. The difference between a 100% charged cell and the depleted cell is only about 0.15 V. Applying a constant voltage across the battery will overcharge or would not charge the battery at all.

Charge termination of the battery is equally important as charging a battery. One major problem with battery charging is early termination. If the charging system is only 95% accurate, then the battery is not fully charged. It is recommended that battery charging should be terminated after detecting $-\Delta V$ or peak voltage, followed by top-off charging.

At this point, a battery can no longer sustain an increase in voltage, because its charge acceptance is low. Pumping in additional energy for charging goes into the production of heat. For this reason, back-up termination is based on an increase in temperature, dT/dt, or a predicted time-to-charge technique. Fast-charge termination is sometimes followed by top-off and trickle charging. The charger electronics detects dT/dt since NiMH batteries do not exhibit a pronounced drop in voltage after reaching near full capacity. A NiMH battery charge cannot be terminated properly if a device monitors $-\Delta V$ only.

The charge sensing electronics can be programmed to support several combinations of detection and termination scenarios. For example, peak voltage detect (PVD) can be enabled such that if a battery reaches PVD before dT/dt, the device terminates charging at a peak voltage. Both the battery charge terminations can be enabled at the same time. The charge electronics uses A/D converters to measure peak voltage to within a 2 mV range. This value is less than 0.6% of the voltage of a battery during charging on a per cell basis.

Alternately, IC makers offer different charge techniques. Charge electronics combines programmable, constant-current based fast charging with overvoltage protection for NiMH batteries. Unlike typical detection methods, such as the $-\Delta V$ or dT/dt methods, the charger controller detects an inflection point d^2V/dt^2. This point is reached by the charged battery at approximately 90% capacity and occurs when the battery voltage increase tends to accelerate. This detection mechanism is NiMH battery friendly as it detects the overcharge process at an early stage.

Upon detection of the inflection point, the charger continues the charge current for another 20-minute period. This is followed by a trickle charge phase to maintain a full charge. In order to prevent an inaccurate voltage measurement, the charging is halted, briefly, while a voltage measurement is taken. In addition, the charge control may include options for automatic predischarge of the battery pack, timed charging, and the choice of use of a switched mode power supply.

CHARGING TECHNOLOGY

With electric vehicles (EVs) comes the EV recharge infrastructure, both for public and private, or domestic use. This infrastructure includes recharging units, ventilation requirements, and electrical safety features suited for both indoor and outdoor charging stations. As an example of the developments, to ensure the safe installation of charging equipment, changes have been made to State of California Building and Electrical codes.

Charging Stations

During EV charging, the charger transforms electricity supplied by the local utility into energy compatible with the vehicle's battery pack voltage requirements. According to the Society of Automotive Engineers (SAE), the complete EV charging system consists of the equipment required to both condition and transfer energy from a constant-frequency, constant-voltage source or network to direct current. The direct current is required for the purpose of charging the battery and/or operating the EV electrical systems (e.g., EV interior preconditioning, traction battery thermal management, onboard vehicle computer). The charger communicates with the battery management system and/or monitor (BMON). The management system and/or BMON in turn calculates how much voltage and current is required to charge the battery system.

Charging is accomplished by passing an electrical current through the battery to reform its active materials into their high-energy charge state. The charging process is basically a reverse of the discharging process. Current is forced to flow back to the traction battery pack. This current initiates a chemical reaction in the opposite direction. The algorithm by which this is achieved differs depending upon the battery type and due to the variations in their chemical compositions.

The EV is connected to the EV supply equipment (EVSE), which, in turn, is connected to the local utility. The National Electrical Code (NEC) defines this equipment as the ungrounded conductors, grounded conductors, equipment grounding conductors, EV couplings and connectors, attachment plugs, and all other fittings, devices, power outlets, or accessories installed specifically for the purpose of delivering energy from the utility wiring to the EV.

For residential or private and most public charging locations, there are two power levels: Level I and Level II. Level I or convenience charging, allows for charging the traction battery pack while the vehicle is connected to a 120 V, 15 A branch circuit. A complete charging cycle takes anywhere from 10 to 15 hours to be completed. This type of charging system uses the common grounded electrical outlets and is used when Level II charging is unavailable. Level II charging takes place while the vehicle is connected to a 240 V, 40 A circuit, dedicated solely for EV traction battery charging purposes only. At the Level II voltage and current levels, a full charge takes from 3 to 8 hours, depending on battery type. In order to sustain the Level II power requirements, EVSE must be hardwired to the premises wiring.

A third power level, Level III is any EVSE with a power rating greater than Level II. Most Level III charging systems are located off the vehicle platforms. Level III charging is defined as the EV equivalent of a commercial gasoline service station. In this case, a Level III charging station can successfully charge an EV in a matter of minutes. To accomplish Level III charging, the equipment must be rated at power levels from 75 to 150 kW. The Level III requires supply circuit to the equipment be rated at 480 V, 3 ϕ and between 90 to 250 A. However, the supply circuit for the Level III charge may be even larger in capacity. The equipment is to be handled by specially trained personnel.

All EV infrastructural equipment, at all power levels, are required to be manufactured and installed in accordance with published standards documents such as: NFPA (NEC Article 625), SAE (J1772, J1773, J2293, others), UL (2202, 2231, 2251, others), IEEE/IEC, FCC (Title 47–Part 15), and several others.

Coupling Types

The EV system will be connected to the vehicle by the general public in all weather conditions. There are currently two primary methods of transferring power to EVs: (1) conductive coupling and (2) inductive coupling.

In the conductive coupling method, connectors use a physical metallic contact to pass electrical energy when they are joined together. Specific EV coupling systems—connectors paired with inlets—have been designed that provide a nonenergized interface to the charger operator. Thus, not only is voltage prevented from being present before the connection is completed, the metallic contacts are completely covered and inaccessible to the operator.

In the inductive coupling method, the coupling system acts as a transformer. AC power is transferred magnetically, or induced, between a primary winding, on the supply side, to a secondary winding, on the vehicle side. This method uses EV infrastructure that converts standard power-line frequency (60 Hz) to high frequency (80,000 to 300,000 Hz), reducing the size of the transformer equipment. The inductive connection is developed primarily for EV applications, though it has been applied to other small appliances.

In both conductive and inductive coupling, the connection process is safe and convenient for all users.

Charging Methods

There are three primary methods of charging EV batteries: (1) constant voltage; (2) constant current; and (3) a combination of the two.

Most EV charging systems use a constant voltage for the initial portion of the charging process, followed by a constant current for the finish. Most of the battery capacity is restored during the constant-voltage portion of the charging cycle. The constant-current portion of the charge cycle, commonly referred to as a trickle charge, serves to slowly top off the battery at a rate sufficiently slow to prevent the off-gassing of either hydrogen or oxygen from the electrolyte.

Building Standards

To ensure that the charging equipment supporting EVs is safe, the National Electric Vehicle Infrastructure Working Council (IWC) was formed to address EV infrastructure issues. The IWC is a consortium of representatives from across the nation and around the world, representing industries, such as electric utilities; automotive engineers; electrical manufacturers; code consultants; EV industry organizations; regulatory agencies; and independent testing laboratories, such as Underwriters Laboratories (UL).

The IWC developed recommended software and electrical code language that addresses the electrical requirements for EV charging equipment; and, along with SAE, submitted code language proposals for inclusion in the 1996 National Electrical Code (NEC). In California, a modified version of these codes has been adopted, which took effect August 19, 1996.

These codes address several issues associated with EV charging equipment. These issues can be classified, primarily, as pertaining to electrical safety devices required in the equipment or the ventilation of the charging system location.

Electrical Safety

Using electrical safety as an example, the EV connector must be polarized and configured so that it is noninterchangeable with other electrical devices such as electric dryers. The method by which the EV charging equipment couples to the EV can be either conductive or inductive, but must be designed so as to prevent against unintentional disconnection. Additionally, the new electrical codes require that EV charging loads be considered continuous; therefore, the premises wiring for the EV charging equipment must be rated at 125% of the charging equipment's maximum load.

All EV charging equipment must have ground-fault circuit interrupter devices for personnel protection, and rain proofing, for outdoor compatible equipment. An interlock to de-energize the equipment in the event of connector or cable damage must be incorporated. Furthermore, a connection interlock is required to ensure that there is a nonenergized interface between the EV charging equipment and the EV until the connector has been fastened to the vehicle.

A ventilation interlock is also required in the EV charging equipment; this interlock enables the EV charging equipment to determine whether a vehicle requires ventilation and whether ventilation is available. If ventilation is included in the system, the ventilation interlock will allow any vehicle to charge. However, if ventilation is not included in the system, the mechanical ventilation interlock will allow vehicles equipped with nongassing batteries to charge, but not vehicles equipped with gassing batteries.

Ventilation

Part II, Uniform Building Code, of California Title 24 Code of Regulations addresses location and ventilation issues associated with EV charging. Specifically, these codes address where EV charging equipment can be installed. If a ventilated charging system is to be installed, the codes specify how much mechanical ventilation must be provided to ensure that any hydrogen off-gassed during charging is maintained at a safe level in the charging area.

The ventilation rates specified in the building codes are calculated to comply with the NFPA requirements published in Standard NFPA 69, Explosion Prevention Systems. This standard establishes requirements to ensure safety with flammable mixtures. Section 3–3, Design and Operating Requirements, requires that combustible gas concentrations be restricted to 25% of the Lower Flammability Limits. This design criterion provides a safety margin in atmospheres containing hydrogen. Hydrogen is combustible in air at levels as low as 4% by volume of air. Therefore, in order for the charging station to not be classified as hazardous, the hydrogen concentration must not exceed 10,000 ppm, which equates to 1% hydrogen by volume of air.

BATTERY PACK CORRECTIVE ACTIONS

Connection Resistance

The intercell and terminal connection resistance is set as a baseline during the installation of the traction battery pack. This recording process ensures that an increase in the resistance values can be detected at an early stage, especially those caused by loose connections and corrosion.

Normal installation resistance varies greatly as a function of the size of the installation. The variation of the installation resistance is typically from 10 to 100 $\mu\Omega$. The installation resistance of a battery or series of batteries can be measured using an ohmmeter or a conductance meter, or a measurement of a voltage drop during capacity testing. It is a good practice to refer to the manufacturer specified guidelines for expected values.

It is a common practice to use either a 20% change in the previously established baseline value or a value exceeding the manufacturer's recommended limit. The corrective actions required are determined by the analyses of the effects of the increased resistance.

Thermal Runaway

When a traction battery is operating on float or overcharge in a fully recombinant mode, the net chemical reaction is minimum. Virtually all the energy ($V \times I$) results in heat generation. During the design of the system, the heat generated during the charge-discharge process should be dissipated without raising the battery temperature. In case the temperature of the battery pack rises during the charge-discharge process, more current will be required to maintain the float voltage. The additional current results in the production of excessive oxygen, which in turn generates more heat. The heat is produced as a result of the reconversion of water at the negative plate of the battery.

The net result is that the battery undergoes a meltdown due to thermal runaway. The possibility of a thermal runaway can be minimized by the use of battery pack ventilation using forced cooling. The ventilation maintains the battery temperature between and around cells. In addition, the charger output (voltage and current) is regulated by using a temperature compensated charge.

Cell/Unit Internal Impedance/Conductance

AC impedance and AC conductance tests are performed to determine battery pack internal impedance or internal conductance of the traction battery. The impedance and conductance tests provide baseline information from similar cells in the battery.

The internal cell impedance of the traction battery cell consists of the physical connection resistances exhibited by battery terminal to battery plate welds and similar plate-to-plate connections. The ionic conductivity of the electrolyte and the activity of the battery during the electrochemical processes occurring at the plate surfaces contribute to the total cell impedance. With the multicell units, there are additional contributions due to intercell connections. The resultant lumped impedance element can be qualified using either AC impedance or AC conductance test methods.

The AC impedance test is performed by passing an AC current of known frequency and amplitude through the battery pack under test. The test accurately measures the resultant AC voltage drop across each cell/unit. The AC voltage measurement is taken between the positive and negative terminals of the individual cells or the smallest group of cells possible. Compute the resultant AC impedance using the Ohm's Law. The ambient temperature, cell and battery life, and discharge history are factors that impact the AC impedance.

The AC conductance test is performed by the user applying an AC voltage of known frequency and amplitude across the battery. In addition, observe the AC current flowing in response to the voltage applied. The AC conductance is the ratio of the AC producing it. Only the in-phase current component is considered for the measurement of AC conductance. The effects of spurious capacitance and inductance impact the out-of-phase component of the battery. The out-of-phase component is not considered for the AC conductance.

AC impedance and AC conductance are inversely related. Cell conductance is directly proportional to the active plate area and decreases as the active area is lost. The loss of active area results through processes of plate aging and deterioration. The decrease in the cell's capacity with time, the cell's AC impedance increases, and its conductance decreases.

When making field measurements of either AC impedance or AC conductance, the observed values should be compared with baseline values appropriate to the specific installation. If such values are unknown, comparisons can be made to values obtained by averaging measurements over the entire string of the battery pack. When the cell impedance increases by more than 30%, it is recommended that the battery manufacturer instructions should be followed. If the cell impedance increases by more than 50%, it is recommended that load analyses of the battery pack should be performed. If the cell impedance increases by more than 80% of its reference value, it is recommended that the battery manufacturer should be contacted for additional actions. Load testing of the battery pack is recommended as soon as the battery cell impedance increases by more than 70%.

Equalizing Charge

During periodic equalization charges, it is not necessary to correct cell/unit imbalances in the battery pack as long as the individual batteries themselves gain the necessary charge. Equalizing the charge should be performed within manufacturer recommended guidelines and limits. This includes the duration of the equalization charge and the magnitude of the current applied during the equalization process.

Ripple Current

In order to limit the ripple current, a battery charger with low electrical noise levels must be used for charging the traction batteries. An acceptable charger is one that does not raise the average fully charged battery pack operating temperature measured at the negative terminal by more than 5°F above ambient temperature in free-standing condition.

Battery Charging Parameters

The charging efficiencies, η_{daily} and $\eta_{overall}$ of the EV can be calculated as follows:

- Determine the miles traveled since the previous battery pack charge
- Determine the kWhr consumed during the recent charge

The daily battery pack charge efficiency

$$\eta_{daily} = \text{Miles Traveled since Last Charge/kWhr Consumed}$$

The overall charging efficiency $\eta_{overall}$ of the EV can be calculated by:

- Determining the miles traveled during the entire test program
- Determining the kWhr consumed during the test program

$$\eta_{overall} = \text{Miles Traveled during the entire Test Program/kWhr used}$$

5 ELECTRIC VEHICLE BATTERY FAST CHARGING

VRLA and NiMH batteries exhibit low internal resistance and are capable of very high discharge and charge rates. This is exhibited by modern battery modules, which are manufactured using thin electrodes with large active areas, allowing for charge and discharge rates of 10C to 20C. Excessive overcharging can on the contrary become catastrophic on these batteries.

THE FAST CHARGING PROCESS

The fast charging technique for traction batteries account for the battery charge acceptance. The charger adjusts the charge rate continually to match the ability of the battery to accept the charge. Danger from excessive overcharging can be avoided, and the battery modules can arrive at the charge in 20 to 30 minutes. This fast charge also enhances the battery life and provides higher battery efficiency (charge recovery).

Battery modules are sensitive to overcharge. The internal resistance of the battery is generally an obstacle to the fast charge and discharge process. Under room temperature conditions the maximum charge/discharge rate of the battery with low internal resistance may be as high as 15C to 30C at room temperature. The charge rate 1C or one-hour rate is the charge/discharge rate in amperes equal to the capacity of the battery in ampere-hours. Similarly, the 10C charge rate is 10 times higher and also referred to as the six-minute rate.

A high-rate discharge of the battery does not cause a catastrophic failure. The battery voltage collapses, and the battery current to the load ceases. A high-rate charge on the other hand can result in serious battery damage. Thus making it important to know when to stop and how to stop.

A fully discharged battery in a 0% SOC can accept current at the highest, initial charge rate. A fully charged battery at 100% SOC cannot be charged any further. Thus passing additional charge current through the battery is meaningless as there is no additional energy available at discharge.

As the battery charging progresses towards a fully charged state, there are a reduced number of carriers available for conversion from the discharged to the charged state within the electrode mass. Thus the ability to accept more charge diminishes continually. The charge acceptance curve describes the maximum charge rate, which the battery is capable to accept (i.e., convert into stored electrochemical energy). Anything above this charge rate constitutes an overcharge. Thus the battery pack can be driven into an overcharge at any time, in any state of charge by excessive charge current.

A previous misconception is that the battery must first undergo a charge before it goes into a charge condition. For example, boiling of lead-acid batteries at the end of the conventional charging, often 25 to 35% of excess overcharge energy is delivered, in order to realize full capacity upon discharge. The reason why the last few percent of the battery capacity take the longest time to charge is because the battery's ability to accept the charge diminishes and tends to zero as the battery pack approaches a full charge.

If the initial charge acceptance ability of the battery pack exceeds the current battery capacity of the charger, 10C as an example, charging will begin at this constant current or "high rate." This area lies below the charge acceptance curve (i.e., in the undercharge zone of the battery). When the charge acceptance region of the curve is intersected, the charge current gradually decreases to match the battery's charge acceptance ability. When full charge is reached, the current is turned off. Alternately, until the battery is removed from the charger, it is rested or maintained at the trickle charge current at the rate of 0.02C or 0.05C. This trickle charge is essential to compensate for the battery self-discharge.

The battery or battery pack being charged using the fast charge process may be used immediately in case the temperature rise during the charge process does not exceed nominal discharge temperature values. In the event the battery or battery pack exhibit a high temperature as a result of the fast charge, it is recommended to allow the battery or the battery pack temperature to stabilize within safe operating values.

The fast charger monitors the battery integrity, battery cell response, high-rate charge, battery controlled charge (battery acceptance and battery matching), finishing charge, and battery charge termination. The fast charger recognizes the boundary of the overcharge zone as shown in Figure 5–1 on the basis of electrochemical potential. The feed-

Figure 5–1   Fast charge profile for traction batteries.

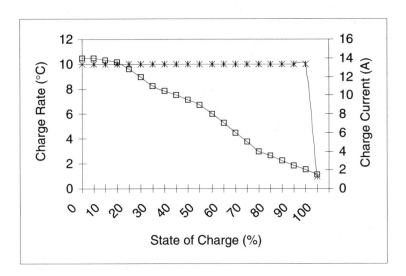

back mechanism allows continuous adjustment of the charger current to match the battery's or battery pack's charge acceptance ability. The fast-charging algorithm also provides battery diagnostics by continuous detection of the electrochemical potential as a function of the battery chemistry and current density. The fast charger provides a high-current density to drive the electrochemical reaction without driving the battery into an overcharge.

Some fast-charging techniques maintain a microprocessor or microcontroller based voltage peak method. The battery pack charge is limited within a predefined charge acceptance curve. This approach is applicable at charge rates up to 1C but cannot be applied successfully at higher battery charge rates. As an example, a 20-minute 3C constant current charge with a conventional battery charger uses voltage peak termination. The pressure rises to 120 psig (10 bar), close to the battery cell's burst limit. This is followed by a sharp rise in the battery temperature and the bursting of safety seals. The heat is produced both during the charging and the discharging process and arises from three main sources. First, as a reversible thermodynamic component, the heat source is the entropy change associated with the temperature dependence of the free energy change of the charge reaction (i.e., with the reversible electrode or cell potentials). The heat produced is $-T\Delta S$, and its rate of production is linear with respect to the battery charge current. The second component is the heat related to the irreversibility of the reaction, which is proportional to the battery overpotential. This is due to the charge transfer process. The third component is the resistive heating, which is proportional to the square of the current and to the cell resistance. In addition, audible hissing is caused along with loss of electrolyte before the battery overpeak voltage is even detected.

In the VRLA battery, heat produced during the charge is determined using the ΔS value as 350 cal/Ahr. In this case, the sign of the heat generated is positive or exothermic in nature. The unit activity of H_2SO_4 occurs at a specific gravity of 1.24 and even at full charge the activities differ from unity by a negligible amount. Most of the sulphate is in the form of bisulphate and only about one-third of the H_2SO_4 is ionized to sulphate ions. Using the bisulphate as part of the reaction, the amount of heat calculated is 53 cal/Ahr — smaller — but again as an exothermic reaction. Reversible heat production occurs at the negative plate, with some cooling at the positive plate. The heat production is exothermic in nature with about 5 mΩ resistance for a six-cell VRLA or AGM battery. The reversible heat produced during the charge is negligible in comparison with the resistive heating. Thus the fast charging process must avoid overcharge and minimize the internal resistance of the batteries.

A fast charger on the other hand will charge even very cold batteries safely. Using a constant resistance-free voltage approach, at a low battery temperature, the entire battery charge acceptance is reduced. The fast charger senses the lower battery charge acceptance and adjusts the charge rate accordingly. The current rises for a few minutes as the battery electrodes come back to life with increase in the cell temperature. In case the charging is interrupted accidentally the battery charge algorithm once again senses the current SOC and reapplies the adjusted charge current. A temperature sensor monitors the battery pack temperature and applies battery temperature compensation over the operating range of the battery charger.

FAST CHARGING STRATEGIES

A large number of charging approaches have been discussed previously. One of the preferred methods by most electrochemists is constant current-constant voltage (CV) interval. This requires some knowledge of the electrochemical processes in order to make a good choice between CV and constant-current (CC) methods.

Using the CC-CV charging method for a fixed high current limit, I_{limit} and for two choices of CV, the current curve for each of the two choices of CV is defined for charging to a low voltage limit, V_L. This is a lower end-of-charge and ensures safe, efficient recharge without significant gas generation. It is assumed that the battery has been subjected to a deep discharge, or heavily sulphated, owing to a high initial resistance in comparison to a fully charged lead-acid battery. Under these conditions, the initial charging current will be less than the current limit set because a substantial portion of the applied overvoltage (V_{app}) will appear as ohmic drop, IR, and relatively little will be available to drive the electrochemical conversion reactions. The voltage will rapidly rise from the initial open-circuit voltage (OCV). If IR is sufficiently large, V_{APP} (lower V_{APP} curve) will quickly reach the low value of limiting voltage chosen, V_L, and the current will continue to rise until it reaches the selected current limit. At constant current, the applied voltage may fall and pass through a minimum limit depending on the value of I_{LIMIT} and whether the resistance is still decreasing. When nonohmic polariztion increases sufficiently, the applied voltage will rise again to maintain the current at I_{LIMIT} until V_{APP} reaches V_L again. Then the current decreases at constant applied voltage to a very, nearly zero, value, as the equilibrium potential increases and the overvoltage is no longer large enough to drive the electrode reactions as shown in Figure 5–2.

Figure 5–2   Fast charge voltage/current profile for VRLA batteries.

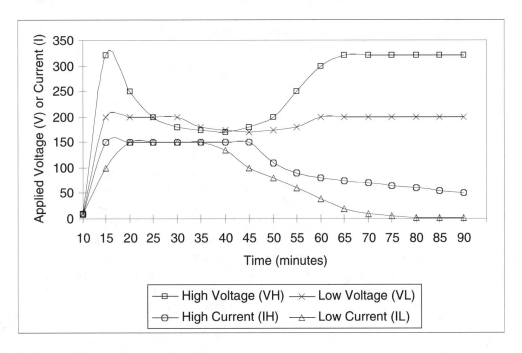

When the applied voltage limit is raised to a high value, V_H, the time at which the battery accepts the maximum current begins earlier and extends longer. When the final current decline occurs, at constant V_H, the gassing reactions remain appreciable, so that the current declines towards a nonzero limiting value. At the low V_L, the charging efficiency is very high, but it may take an impractically long time to return all the charge because of the very low current. This phenomenon is illustrated by the already low current when 80% SOC is reached. At the high voltage, the current remains high. The average rate of charging is faster but less efficient, and a high proportion of the charge goes into gassing reactions—particularly after about 80% SOC. Eighty percent SOC is sufficient to reduce the value of the applied voltage to V_L if 100% return is not needed. In order to attain the intermediate value, the voltage is tracked using artificial intelligence. For automatic charging, a computer algorithm determines the SOC with user specified voltage and current limits.

As an alternative, feedback control is employed so that the ohmic component of the voltage decreases. The user must still specify the current limit and a resistance free voltage, V_{FR}—but this, instead of the applied voltage, is maintained constant.

When a constant resistance-free voltage is applied during the fast charging, the initially applied voltage rises to a value limited only by the power supply. After the first few seconds, most of the overpotential is ohmic, and the current reaches I_{LIMIT}. As the resistance decreases, the applied voltage decreases until other nonohmic polarizations (η_{NOP}) become important at about a 40 or 50% SOC. The applied voltage V_{APP}, rises and maintains $V_{APP} = V_{RF} + IR$ from the point the current begins to fall. Then V_{APP} must also fall since V_{RF} is constant. By choosing the applied voltage suitably, optimum charging rate is obtained, since the current remains at its maximum value for the longest time consistent with the desired charging efficiency and temperature rises.

The optimum value of V_{RF} is not truly a constant but varies according to the acid concentration, kinetic and thermal characteristics of each battery in the pack. It is therefore a function of the SOC of the battery. The principle of charge control used here is an approximation to the ideal battery-charging algorithm. The algorithm accounts for the factors influencing the battery pack performance and compensates for them.

For an 85 to 90 Ahr battery pack, a fast charge is applied at 8C to 9C. The current limit is set to 450 A and is somewhat lower, and the value of V_{RF} is 2.50 V. The temperature compensation for the V_{RF} is set to 6 mV/°C/cell.

Both the charging voltage peaks are noticed when initial decline is caused by the decreasing battery resistance during the CC charge interval, with the second decline due to decreasing current. In conclusion of the fast charge it can be ascertained that:

- The commercially available batteries can be fast charged with a temperature rise of less than 25°C
- Fast charging does not exhibit detrimental effects on battery cycle life

In the first part of the fast charging process, to about 40% return, resistive heating is the major cause of the initial battery pack temperature rise. In case the battery is designed such that it meets the five-minute-50% charge return requirement, the best strategy is to design the battery pack with low resistance batteries.

The temperature distribution of batteries in the pack appears nonuniform, with possibly temporary localized hot spots. In the case of a VRLA battery, the temperature on the casing surface follows the internal temperatures more closely during charging than for immobilized electrolyte batteries.

A battery with a larger heat capacity will have a lower temperature rise when it is subjected to fast charging. Assuming that all other battery pack conditions are the same. Therefore, VRLA batteries have a lower temperature rise, since sulphuric acid solutions have a large heat capacity. VRLA batteries will also have less heat production from water vapor decomposition. In addition, the battery heat transfer to the battery exterior is better.

The lowest charging current should be chosen such that it takes full advantage of the time allowed to meet the charging requirement. This lowers the resulting battery temperature and saves energy. In addition, it may also return more charge to the battery in the given time.

THE FAST CHARGER CONFIGURATION

The fast charger for traction batteries provides charging of batteries in 5 to 30 minutes. In order to apply this fast charge in a period of 10 to 30 minutes, the charger must be able to provide voltages up to 450 V and currents up to 500 A. Such a charger characteristic "envelope" is depicted by a maximum voltage-maximum current profile as shown in Figure 5–3. This envelope implies a peak output power of 225 kW. Thus

Figure 5–3 Constant voltage charging profile.

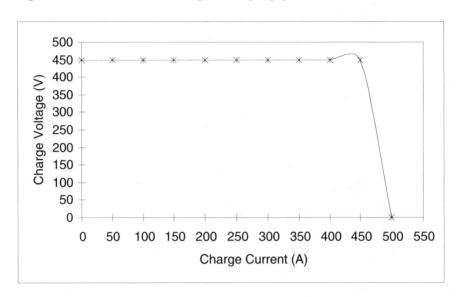

charging a compact electric car in about 6 to 10 minutes, a midsize electric vehicle (EV) in about 25 to 30 minutes provided that the battery quality allows charge acceptance at such rates.

The modified battery charge profile of maximum voltage-maximum power-maximum current limits the power to 120 kW, while the maximum current and the maximum voltage characteristics remain the same as the maximum voltage-maximum current envelope, as shown in Figure 5–4.

Using this new charge envelope, the compact size EV will require 10 to 12 minutes to charge. The maximum voltage-maximum power-maximum current profile characteristic has the advantages of (a) lower peak power (which dictates that the size of the battery grid); and (b) spread of charging times, among EVs can be narrowed down in spite of different battery voltages and battery capacities. The charger is designed essentially to deliver the same amount of energy in the same time.

The fast charger requires intimate knowledge of the battery on charge. The battery charger requires knowledge about battery pack, and the faster the charger, the more is the information needed. The charger can prevent unwanted abuse of the battery while achieving optimal charge in the shortest possible time. Such information includes the battery chemistry, number of cells in the battery module, and voltage and temperature characteristics of the battery. The charger control can also

Figure 5–4   Maximum voltage-maximum power-maximum current charging profile.

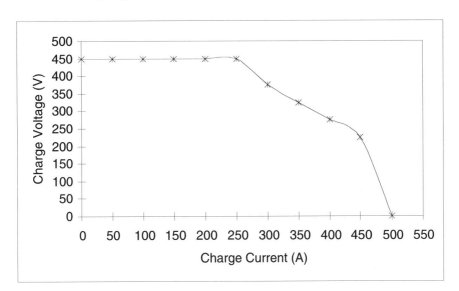

become more "battery specific" on an individual battery basis and monitor the battery under charge.

The fast charger defines the universal charging station for EVs. The salient features are based on the battery charger, comprised of the power section and the controller section.

The power section of the battery charger is a DC-controlled current power source with sufficient voltage compliance to be able to charge a wide range of EVs, not being battery specific. The controller section of the charger is battery specific and is placed on board the EV together with the battery pack.

The control signal connection (a twisted pair or coaxial cable) is required for the battery pack charge control. Additional wires are required to send a control signal for the automatic mode operation, and transmit the battery pack temperature and battery charge acceptance data to the battery monitor (BMON) module. The power cable along with the control pair constitutes the charging cable interface. A standard high-voltage connector interfaces the charging station to the EV battery pack charge receptacle.

A large range of EVs can be charged as long they are equipped with a standard charge receptacle. In such a configuration, the charge is performed in the manual mode. The user provides the voltage and the current level, within the battery-charging envelope. In addition, the user may also input the charge time and the total Ahr to be provided to the battery pack. This type of control places limits on the output current and the output voltage of the battery pack, allowing the battery pack to be charged using the popular constant current–constant voltage charge profile. In case the values for the charge current and the charge voltage are not available, the user should consult the manufacturer specifications.

The high-level components of the EV fast battery charger include the charging station interface consisting of an electrical connection to the power supply grid capable of supplying several hundred amperes. A rectifier module converts AC input power to DC input power. A switching inverter module regulates the flow of DC charge power to the battery. The charge delivery cable carries the power to the EV battery pack. The termination of the charge delivery cable is a receptacle jack, which carries hundreds of amperes of DC current. This receptacle may require substantial insertion force to achieve a reliable connection to the battery pack. It is thus necessary to provide a locking lever that will assist with the insertion and the removal of the battery charge cable. Such a zero insertion force design will be required for public acceptance. Thus making the insertion and the removal of the plug as simple as the gasoline nozzle used to fuel the internal combustion engines. A connector-locking lever will also ensure that the control signal is activated upon proper contact before high current is delivered through the battery charge cable.

The station controller regulates the charge current either in the (a) automatic mode, as a slave to the charge controller on board the vehicle; or (b) in manual mode, according to the manual current, voltage and time settings. The switching inverter and the station controller are designed in such a way that they are capable of fast turn-on and turn-off in several milliseconds. A digital readout provides indication for charge voltage (V), charge current (A), delivered energy (kWhr), and the monetary charge in the currency units ($). The EV is equipped with a jack, a battery pack, and a charge controller. The charge controller monitors the battery pack (or its individual modules) by sense control lines. The charge controller communicates with the battery station controller via the signal lines.

Fast Charging Prerequisites

Fast charging prerequisites include:

- All personnel performing fast charging of the battery pack must observe proper safety precautions at all times.

- Charging rates will not exceed maximum battery charging rates.
- If charging of the battery pack is immediately followed by a fast charge, the time interval sequences must be as per the EV owner's manual.
- The ambient temperature of the battery pack must not exceed 120°F at the start of the fast-charge cycle or the maximum temperature allowed by the battery manufacturer, whichever is less.
- Fast charging at 120 VAC may be carried out to meet the manufacturer requirements. However charging at this voltage takes too long.
- The battery manufacturers provide a charger that will fully charge a battery pack from any state of discharge in less than 12 hours.

When the charge has been completed, the following battery pack parameters are noted:

- Charge time
- End-of-charge battery voltage
- Final charging current
- Charging station location energy meter reading
- Vehicle kWhr reading (in case a Whr meter is installed)
- Totalizer meter or data logger information
- Vehicle odometer reading

USING EQUALIZING/LEVELING CHARGERS

Fast charging of the traction batteries returns a large amount of energy to the traction batteries in a short period of time. However, the battery still requires periodic equalizing charges. The battery manufacturer is responsible for determining the frequency of equalizing the propulsion batteries.

To ensure that the equalization charge is completed within the period specified by the manufacturer, the user should complete equalizing charges at the charge intervals specified by the manufacturer.

Limitations of Fast Charging

Maximum charge and discharge rates are both functions of battery design. In addition, traction batteries also behave symmetry while responding to the charge and the discharge process. The battery capable of a high power discharge in 10 minutes is also capable of a high power

charge in 10 minutes. Similarly, a traction battery best designed to
deliver its total energy over a period of one or two hours will need one
or two hours to achieve a full charge. In case of an internal combustion
engine refuelling, the filling tube and the flow control represents a constant rate restriction. While recharging the traction battery may be performed at a fast rate, the process is completed without abuse as a variable
rate process. As the charging of the battery pack progresses, the ability
of the battery pack to accept the charge at the same rate diminishes and
tends towards zero. Thus it becomes more and more difficult to convert
the applied charge current to stored electrochemical energy. Thus the
last 5 to 10% of charge and discharge takes the longest time as shown
in Figure 5–5. It is possible to force the charge current through the
battery pack at a higher rate, exceeding the battery pack charge acceptance specifications. However, this does not result in faster battery charging. Rather it leads to battery abuse and failure due to excessive gassing
and generation of heat.

Thus, it is important that fast chargers are capable of sensing the
battery SOC and also adjusting the applied charge current based on the
ability of the battery to accept charge. This results in the fastest possible charge with a minimal waste through battery overcharge and
minimum battery heating.

**Figure 5–5   Variation of battery charge current and battery
capacity.**

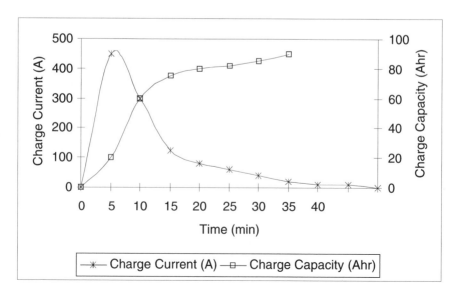

Fast Charging and Battery Overcharge

A fast battery is measured by exhibiting a low internal resistance. Thus a small amount of charge overvoltage is needed to acquire a substantial charge rate C. Both the electrochemical component of the overvoltage and the electrical component must be sufficiently low to allow fast charging while generating only a small amount of heat. In other words the battery can be driven into the charge and/or discharge reaction easily, and the resistance of the conductors and the interfacial contacts can be easily overcome. Thus overcharging of the fast battery during fast charging should be avoided. A large amount of irreversible heat generated in the process will cause permanent battery damage or destroy the battery pack in a very short duration.

During the fast charging process of a 90 Ahr battery, the charging proceeds at a rate of 5 to 8C (450 to 720A) under charger limited conditions. During the application of this charge the battery receives the first 50% of its charge during the initial 10 minutes. As soon the battery charge acceptance drops below 450A, the charger begins to reduce the applied current. The 75% charge is achieved in the next 5 minutes (15 minutes), 90% charge is achieved in the next 7 minutes (17 minutes), 95% of the charge is achieved in the next 12 minutes (22 minutes) and the full battery charge is achieved in the next 25 minutes (35 minutes).

An interesting observation is that the first 10% charge is accepted in less than 2 minutes. The last 10% charge is accepted in 20 minutes. Thus taking 10 times longer to accept the same amount of charge but at a later time and charge condition. The first 5% charge is accepted in less than 1 minute or 60 seconds. The last 5% of the charge takes 15 minutes or 15 times longer than the same amount of charge but at a later time and charge condition.

This explanation is necessary to illustrate the battery pack behavior and the impossibility of delivering the last portion of the fast charge to the battery pack. In case of the EV charging process, it is neither necessary nor practical to wait for the last 5, 10, or even 15% of the battery charge. This is analogous to waiting for a long time to fill up the last gallon or two. Since partial charges are not detrimental, within battery tolerances, it is recommended to apply an equalization charge to the battery pack once in every 20 to 30 charge–discharge cycles. In this case, a conventional slow charge is applied as an equalization charge.

Overcharge of the battery is not only wasteful, but also rather bad for the battery pack. Especially, overcharging of the battery is probably the single most important factor in the reduction of the useful battery life.

The overcharge does not dispute usefulness or necessity of an occasional extended equalization cycle. The equalization cycle brings all the cells in a battery pack in step or balanced. The equalization cycle also ensures that uniform charge distribution is available for the electrode active material. The overcharge energy drives the battery into electrolysis, resulting in elevated potentials, gases, and increase in battery temperature. In the case of vented battery modules, the overcharge results in heat loss and water. During the rapid charging process, the gases recombine, and unless the recombination rate is insufficient, the cells overheat and burst—thus converting the overcharge energy into harmful heat.

During the rapid charger design, it is important to account for a second temperature-sensing device in the battery tub. In the event of failure of the first temperature-sensing device, the second sensor can continue monitoring the battery pack temperature providing thermal protection. In case a second device is not included in the battery pack design, rapid overcharge can result in the failure of the battery pack.

Fast Charging and Battery Degradation

In the event of infrequent charging, traction batteries need a rapid battery charging capability. Rapid charging implies a rapid reverse electrochemical reaction that depends on both battery chemistry and geometry. A high current induces a quick charging reaction, which in turn results in reducing the battery efficiency of the charging process and creates overheating. Maintaining a high charging current can also result in irreversible degradation of the battery due to electrolyte leakage and crystalline dendrite formation.

During the discharge process, current flowing from the battery is determined by the rate of charge transfer resulting from the reduction-oxidation (redox) reactions at the battery cell plates. This rate of reaction is constrained by the amount of electrolyte that is in contact with the electrode. During the charging process, the level of current flowing into the battery is controlled by the electrical battery charging system and may be in excess of the natural redox reaction capacity of the battery cell electrodes. Excess-charging current results in electrolysis of the aqueous electrolyte away from the electrodes and not in the charging of the battery. In the case of a VRLA battery, this electrolysis frees gaseous hydrogen and oxygen instead of soluble hydrogen and oxygen ions. When gas bubbles come in contact with the electrode surfaces, they block the access of the electrolyte ions to the surfaces. This reduces the effective electrode surface area and diminishes the current handling capability of the battery during the charging process.

Since the gas buildup displaces more volume than a liquid solution does, an improper electrochemical reaction produces increases in the pressure and the cell temperature inside the sealed battery cells. When the pressure build-up inside the cells exceeds beyond the VRLA battery seal capacity of the cell case, it results in the leakage of hydrogen and oxygen gas with possible leakage of the aqueous electrolyte solution. Pressure-relief venting of the battery gases mitigates case damages and leakage, but any form of fluid loss lowers the volume of the cell electrolyte available for electrode immersion. Lowering the electrolyte volume decreases the area of the electrode/electrolyte interface. Fluid loss from the battery decreases the current-handling capacity of the battery and is the cause for a large number of battery failures.

Overcharging of the VRLA battery can result in the growth of dendrites or crystalline fingers. This growth is initiated by (1) a charging process after a partial discharge of the battery; (2) the over-charging of a certain type of battery cell; and (3) a prolonged, low-current "trickle" charge process.

A partial discharge followed by a charge leaves some crystals on the cell plate. A uniform distribution of newly formed crystals on the plate implies that there is clumping of new crystals with the already present crystals on the plate. Repetition of this clumping process results in the fingerlike growth of larger crystals known as dendrites. During the discharge process, the crystals do not dissolve entirely due to a small surface-to-volume ratio. In the event that the crystals do not dissolve in some batteries, this results in the so-called "memory" effect and the cell's full-current charge capacity will appear to drop. This effect increases as the dendrites grow larger and cover an even larger area. Depth of discharge (DOD) prior to the charging has an effect on the maintenance of a battery's effective charge capacity.

In the VRLA batteries, the charging process is a reduction reaction that transforms the soluble zincate ions in the electrolyte into crystalline zinc. This process ideally occurs on the surface of the zinc electrode or in its pores. Overcharging of the battery implies continuation of the reduction after all the exposed electrode surface area has already been covered with crystals. The continued reduction of the extra zincate in the electrolyte results in the precipitation of zinc crystals within the electrolyte and not the electrode. This continuous reduction of the zincate and precipitation of the crystals leads to the growth of crystals on the separator pores.

Trickle charging is the most common method of both improving the efficiency of the charging process and also of avoiding overcharging. A charging current that is too low results in electric potential fields that are too weak to maintain the diffusion process. Ions that are not diffusing through the electrode surface collect at the surface. This process, over a period of time, results in a build-up of dendrites that only further inhibits diffusion into the pores.

Large intervals of dendrite growth result in the growth of crystals through an insulating separator. This leads to the bridging of the cell electrodes by the crystal. Thus batteries that are prone to dendrite growth must be stored in a discharged state. The discharged battery must be charged at the maximum current that does not result in excessive gassing or heating. The charge is continued until the level of charge reaches that required by the battery application.

Fast Charging and the Electrical Utility

While the EV is at home in the garage, it is convenient to use the overnight charge. However in a fleet parking garage, the overnight charge will make the overnight charge unprofitable. Especially when the gasoline driven vans run for 5 to 7 hours and take 10 minutes to refuel, the EVs run for an hour or two and take 10 hours to fill up.

From the electrical utility standpoint, the total energy required during battery pack charging is the same when the vehicles are charged slowly, in parallel, or whether they come in to the charging stations sequentially. However, this charging is done during peak load hours. In the case of home overnight charging, the EVs are consuming energy in the off-peak hours. In some countries, electricity is primarily generated for light duties and for home lighting purposes only. In case extra kilowatts of energy consumption are added at home, the circuit breakers begin to reset and the fuses open.

With the improving EV efficiency, the shift away from the fossil fuels will greatly benefit the environment. However if every household has one EV being charged daily, the energy consumption of the power supply grid will double.

In fact the infrastructure that is in place currently allows for a better power distribution to charging stations at selected locations in multiples of megawatts, than in individual homes garage in the country.

In case the rapid charging is to be successful without battery pack failures, it is important for the batteries to tolerate an initial 15C charge rate providing the battery pack with 90% of the charge in about five minutes.

The power to the charging station will have to be made available at multitiered prices. During the off-peak hours, the cost will be lucrative to encourage users of EVs to charge the vehicles. During the shoulder or in-between peak and off-peak hours, the price will be higher. And during the peak hours, the price for charging the EV will carry an additional premium to provide charging opportunities to users in a must have situation.

INDUCTIVE CHARGING—MAKING RECHARGING EASIER

As the demand for the EVs continues to rise, it is important to make it easy to charge the EV. The recharging of consumer and commercial EVs require:

- Simple, easy to use, and intuitive methods
- Flexible infrastructure for all vehicle and utility power levels
- Safe for all-weather operations
- Reliable and long-lasting or durable equipment operation

Most EVs will be charged at least once during the day. This will average out to 3,000 to 7,000 charge cycles during the 10-year life of the vehicle. The conventional electrical outlet cannot meet the heavy-duty cycle requirement of the EV battery pack. During the inadvertent connecting and disconnecting from and to the charge port, large electrical arcs may be drawn, which can be fatal at times. It is essential to use a safe coupling method to charge the EV and prevent all forms of direct electric connection between the electric utility power outlet and the battery pack inlet port. In addition, it is essential to ensure that there are no moving parts associated with the charging mechanism. The charge coupler should offer minimum contact resistance and be rugged enough to withstand the weight of an EV, resistant to the weather elements.

Inductive charging is a coupling process that transfers electrical energy from the electric utility charge port to the EV battery pack through an electromagnetic connection rather than physical or direct connection. Operating on the principle of a transformer, the electrical energy transfer takes place by linking the electromagnetic fields between two separate inductors. The primary and the secondary inductor are coils of conductive wire that are wound to contain the magnetic field with the ferrite material.

A simple inductive coupler consists of a copper coil wound around a ferrite core to direct the magnetic field. When electric current flows through the primary coil, the resulting magnetic field induces an alternating voltage through the magnetic field and into the secondary coil. Thus the circuit is completed. The AC current is then converted to DC and stored in the battery pack.

During the inductive charging process, the onboard vehicle charging port closes mechanically after the coupler is inserted. Thus there are no moving parts associated with the charging process. Since the inductive charger serves as an isolation transformer, it provides electrical isolation between the EV and the utility grid. In addition, the coupler has a double-insulated coil. Both the inductive coupling and the power cable are electrically cold. Power is applied only when the coupler is inserted and locked into the receptacle port, and communication between the EV charge inlet port and the utility outlet port has been established. In addition, the necessary diagnostic tests must be completed successfully, before the power is applied to the cable and the coupler.

Since there is no direct electrical contact, the inductive coupler/port can be operated safely under water. The charge coupler and the charge port are coated with nonconductive protective material and so no water, dirt, ice, or snow will affect the unit or interrupt the safe operation of the inductive charger. Since the orientation is intuitive, there are no additional alignment requirements. The inductive coupler is lightweight and easy to handle. Conventional industrial electrical cables and connectors are heavy, having complex key insertion systems and require an insertion force of 20 to 30 Nmtr and a high twist-to-lock force of 5 to 10 Nmtr.

The one-size-fits-all connector for different utility service lines and battery voltages allows for easy charging. A single-sized inductive coupler with the high-frequency switching electronics can accommodate charging levels ranging from 1 to 150 kW power levels. Thus the inductive charger and coupler with high frequency switching electronics can accommodate charging over the power levels from 1 to 150 kW. This makes it possible for consumers to charge the battery pack at home and at work with a 240 V AC, 6.6 kW system. This quick charge takes place in 12 minutes from 80% SOC at 50 kW from a 600 V AC industrial service line or charge at 1.5 kW with an onboard charger from any 120 V AC outlet. Power is transferred through an inductive system at high frequencies (80 to 300 kHz), making it possible for the coupler and port to be kept at the same size.

RANGE TESTING OF ELECTRIC VEHICLES USING FAST CHARGING

The constant speed range of electric vehicles (EVs) is tested using ETA-TP004 and SAE J227a test profiles. However, using rapid charging, the EV can be charged in a short time. But in order to successfully perform range tests on the EV, there is a set of initial conditions that must be satisfied:

- Road tests are performed on a test track, which is level to within 1%.
- Initial battery pack temperatures during road testing will be within the range of 60°F and 120°F.
- Ambient temperature during road testing will be within the range of 40°F and 100°F.
- The recorded wind speed at the test location during a test will not exceed 10 mph.
- Vehicles are tested at curb weight plus 332 lbs. Additional instrumentation will affect the test weight and the balance of the EV.
- Tires being used will be inflated to the manufacturer's recommendations under cold temperature conditions.
- All manufacturer-recommended lubricants are employed during the tests.
- EV accessories are not used during the test activities.
- All the range-related tests commence with batteries charged to the normal shut-off point of the fast charger.

The range test determines the maximum driving range achieved during a 12-hour period when the EV is operated over a three-day consecutive driving period. The test schedule varies on each day and the driving time, miles driven in a 12-hour period, and kWhr-AC consumed during charging are all noted for the duration of the test. (Refer to Appendix C for a Range Test Log). The total miles driven by the EV over the three-day period is averaged to determine the average miles for each day.

Driving Range at the End of Day-1 Test

The vehicle driving range test is to determine the maximum driving range achieved at the end of Day-1 during a 12-hour period. During this test period, the battery pack is periodically charged using the rapid charger. The minimum range traversed by the vehicle in an 8-hour interval is 100 miles. With a 332 lbs passenger or equivalent weight, the vehicle travels 35 miles between any two charging segments.

During this driving test, the values noted include:

- Average vehicle speed
- Average distance required between charge
- Average distance traveled between charge
- Average kWhr available per charge

Driving Range at the End of Day-2 Test

The vehicle driving range test is to determine the maximum driving range achieved at the end of Day-2 during a 12-hour period. The test is repeated with the same passenger weight and minimum range requirements as on Day-1. The same driving test values are noted during the test on Day-2.

Driving Range at the End of Day-3 Test

The test is repeated with the same passenger weight and minimum range requirements as on Days-1 and -2. The same driving test values are noted during the test on Day-3.

ELECTRIC VEHICLE SPEEDOMETER CALIBRATION

In the event the EV speedometer requires recalibration, the vehicle requires a data acquisition system.

The speedometer can be recalibrated using the following steps:

- Record the speedometer reading when the EV is stopped.
- Accelerate the EV to 5 mph and record the speedometer reading.
- Increase the EV speed in increments of 5 mph and note the speedometer readings until the final speed of 60 mph is achieved.
- Increase the EV speed to 80 mph and note the speedometer reading.
- Develop a speedometer calibration table, taking differences in the speedometer reading and the calculated correction factors into account.
- Using this speedometer calibration table, repeat the test with a re-calibrated speedometer.

6 ELECTRIC VEHICLE BATTERY DISCHARGING

The U.S. Advanced Battery Consortium (USABC) has prepared battery test information that provides test guidelines for full-voltage battery packs, battery modules, and battery cells. The test procedures provide procedures and parameter values to be used by all in evaluating the batteries, including the battery developers and other test facilities.

The Variable Power discharge test or the simplified version of the Federal Urban Driving Schedule (FUDS) was developed by the Department of Energy/Electric-Hybrid Vehicle Program (DOE/EHP) Battery Test Task Force (BTTF) in 1988 provides an effective simulation of the dynamic discharge conditions (for driving-cycle testing) in the laboratory. The simplified version profile was modified into the Dynamic Stress Test (DST). The DST is scaled to a percentage of the maximum rated power or USABC power goal, and requires higher regeneration levels than the SFUDS cycle. Table 6–1 summarizes the corresponding power values. The 100% power value is intended to be 80% of the USABC peak power goal for the technology. For example, if this profile is scaled to 80% of 150 W/kg, it would have a peak power of 120 W/kg.

Commencing with a fully charged battery, the battery is discharged by applying the scaled DST power profile. The 360-second discharge test is repeated with minimal time delay (rest period) between the discharge profiles until the end-of-discharge point specified in the test plan or the battery voltage limit, whichever occurs first, is reached. The end-of-discharge point is based on the net battery capacity removed (total Ahr-regeneration Ahr). In addition, the DST test provides insight into the VRLA battery's changing internal resistance simulating the dynamic driving conditions of the electric vehicle (EV).

The DST test simulates the dynamic driving conditions that EV batteries may see.

In addition, to performing these tests at room temperature conditions, effects of temperature may be simulated by maintaining the batteries at several different temperatures including –10°C, 25°C, and 50°C.

Table 6–1 Listing of DST power profile.

Step	Duration (secs)	Cumulated Duration (secs)	Discharge Power (%)
1	16	16	0
2	28	44	−12.5
3	12	56	−25
4	8	64	12.5
5	16	80	0
6	24	104	−12.5
7	12	116	−25
8	8	124	12.5
9	16	140	0
10	24	164	−12.5
11	12	176	−25
12	8	184	12.5
13	16	200	0
14	36	236	−12.5
15	8	244	−100
16	24	268	12.5
17	8	276	25
18	32	308	−25
19	8	316	50
20	44	360	0

In addition, the internal resistance values may be calculated for the I v/s V measurements taken during the DST test from steps 14 and 15. The termination voltage of the DST test is based on a cutoff-point at which the battery voltage reaches 9 volts. As the temperature decreases, the number of DST frames completed before the termination decreases as shown in the figures above and vice versa as the temperature rises, the DST frames completed increases. It may also be noted that the R_{int} at the second data point (near the 100% SOC) is very close to the R_{int} (near the 100% DOD).

As the VRLA battery undergoes discharge cycles, its capacity changes with temperature. Although temperature effects on single cells and battery modules are well researched, the performance of several batteries connected together has not been thoroughly researched. In addition to the charging, recharging of battery packs is important to maintain a uniform temperature in the battery pack. Maintaining a uniform temperature is vital both for the battery life and the EV performance. Unfor-

Figure 6–1 Dynamic driving battery discharge test profile.

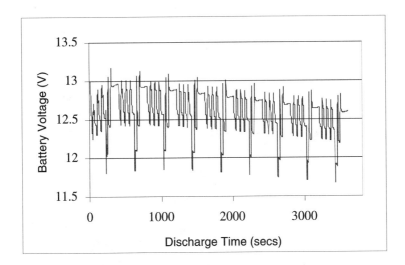

tunately, it is difficult to keep the battery pack temperature, housing hundreds of cells at a constant temperature. This is owing to the fact that the batteries that are on the outermost edges have a greater surface for heat exchange while the batteries in the middle of the pack have the least available surface for exchange of heat. The stacking arrangement of the batteries in the pack leads to nonuniform temperatures, which in turn lead to nonuniform discharge and charge characteristics of the battery pack.

DEFINITION OF VRLA BATTERY CAPACITY

As part of a battery pack configuration, a major problem experienced with the EVs is the premature decline of battery capacity, which ultimately leads to battery failure. The primary cause of the battery pack failure is owing to repeated nonuniform discharging and charging of the cells. Both the battery charging and discharging are highly dependent upon temperature. Owing to the large temperature difference between the coolest battery or batteries on the outer edge of the battery pack, and the hottest battery or batteries on the inner side of the battery pack, there is a corresponding variation in the available battery discharge capacity.

Figure 6–2 Variation of battery capacity with respect to discharge temperature.

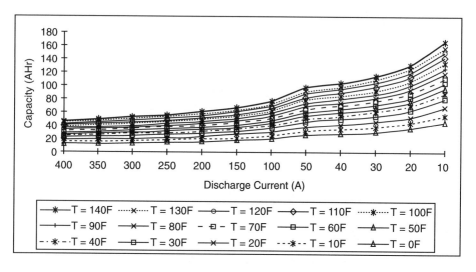

The battery pack cut-off point is normally determined by a predefined total pack voltage. This pack voltage is with respect to a particular current. As the battery pack undergoes discharge, the coolest cells with less available capacity are discharged further to a lower state of charge (SOC) than their hotter counterparts. During the discharge, the capacity of the coolest batteries may be reduced enough to force them into reversal of polarity (i.e., the batteries are reverse charged while the other batteries in the pack are discharged normally). Gradually these repeated over-discharges reduce the battery life. Equalization charges applied to the battery pack balance the batteries. However, the variations in temperature bring back the same discharge and charge limitations as before the equalization.

The temperature gradient affects larger battery packs more significantly, resulting in capacity imbalance and charge acceptance problems that lead to early battery failures. The capacity (C_T) of the battery pack is determined on a per cell basis using a linear two-hour discharge curve based on the equation below. This equation may overestimate the capacity at extreme temperatures.

$$C_T = C_{30} \times [1 + 0.008 \times (T - 30)] \text{ where T is the temperature in °F.}$$

Figure 6–3   Variation of load voltage at varying battery capacities.

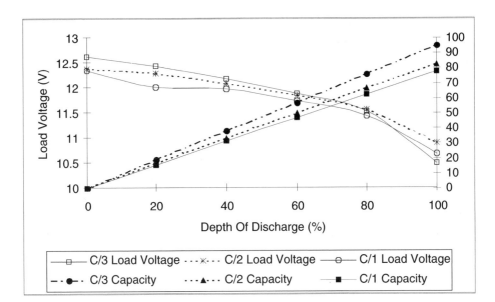

In addition, a variation of temperature and its effect on battery capacity for an 85 AHr battery results in a capacity spread to less than 40 AHr at 0°C. Assume that for a 50 A discharge the highest battery temperature in the pack is 38°C and the lowest battery temperature in the pack is 20°C, the corresponding discharge capacity drops from 52 AHr to 40 AHr. This large difference in the available capacity can cause the lower-capacity batteries to overdischarge and possibly reverse in case most of the batteries are at elevated temperatures and the total pack voltage is used to determine the battery pack cut-off voltage.

Figure 6–3 illustrates the variation of useful battery capacity and battery voltage with changing current discharges at 80°F.

DEFINITION OF NIMH BATTERY CAPACITY

NiMH batteries are also rated with an abbreviation C, the capacity in Ah. The C rating is obtained by the NiMH battery by thorough conditioning of the individual NiMH cells. This is done by the user subjecting the cell to a constant-current discharge under room temperature.

Since the cell capacity varies inversely with the discharge rate, capacity ratings depend on the discharge rate used during the discharge process.

For NiMH batteries, the rated capacity is normally determined at a discharge rate that fully depletes the cell voltage in five hours. For the purpose of electrical analysis of the battery cell, the Thevenin equivalent circuit is used. This circuit models the circuit as a series combination of the voltage source (E_0), a series resistance (R_h = the effective instantaneous resistance), and the parallel combination of a capacitor (C_p = the effective parallel capacitance) and the resistor (R_d = the effective delayed resistance).

Under steady state conditions, the cell voltage at a known current draw is $E_0 - iR_e$, where Re is the effective internal resistance of the NiMH cell. R_e is the sum of the R_h and R_d. Under transient discharge conditions, as shown in Figure 6–4 the initial voltage drops immediately to $E_0 - iR_eh$ and then transfers exponentially, with time constant $C_p \times R_d$ to a steady state voltage. This discharge condition reverses once the load being applied is removed from the battery as seen in Figure 6–4. Note that the slow recovery of cell voltage after removal of the load after 11 minutes is attributed to the delayed resistance R_d. This behavior is identical to the effect noticed during discharge between 4 to 11 minutes.

Figure 6–4   Variation of battery discharge voltage.

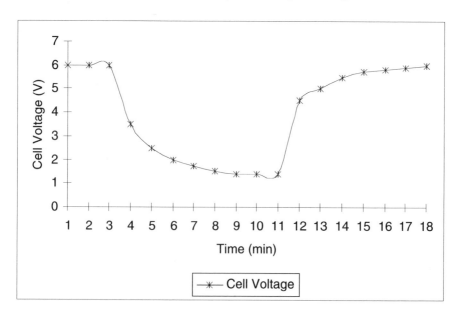

For most applications, unlike EV applications, the steady state voltage is adequate for describing the battery performance. This is owing to the fact that the time constant for most cells is small—typically less than 3% of the discharge time. Although the instantaneous resistance of the NiMH cell is comparable with NiCd cell, the delayed resistance is 10% higher. For this reason, the steady-state voltage for the NiMH cell is lower than that of NiCd.

Voltage During Discharge

The discharge voltage profile for an NiMH cell is affected by transient effects, discharge temperature, and discharge rate. Under most conditions, the voltage curve retains the flat plateau before a rapid drop off termed as the knee of discharge curve as observed between 80% and 100% discharge.

A typical discharge profile for a cell discharged at a five-hour rate (0.2 C) is shown in Figure 6–5. The initial open circuit voltage drops from 1.4 to 1.2 volts. This drop occurs rather rapidly. As seen by the flatness of the plateau and the symmetry of the curve, the midpoint voltage (MPV—the voltage when 50% of the available cell capacity is depleted

Figure 6–5   NiMH battery cell discharge profile.

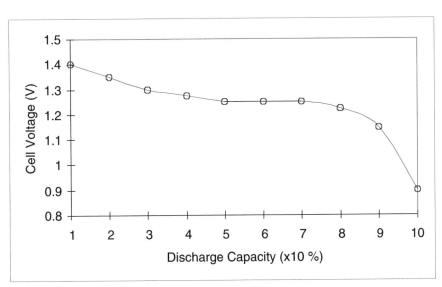

during discharge) provides a useful approximation to the average voltage available throughout the discharge cycle.

Effect of Temperature on Discharge

As noted earlier, the main environmental influences on the location and shape of the voltage profile are discharge temperature and rate of discharge.

Small variations from the room temperature do not affect the nickel metal hydride cell voltage profile. However, large deviations especially of lower temperatures, reduce the midpoint voltage of the cell while maintaining the general shape of the voltage profile. Thus resulting in a diminished useful capacity of the NiMH battery. (See Figure 6–6.)

Discharge Rate

When the traction battery undergoes discharge, the rate of discharge does not have significant effect on the shape of the voltage profile of the NiMH battery under 1C rates. However, for different higher discharge rates, both the beginning and the ending transients consume a larger portion of the discharge durations. (See Figure 6–7.)

Figure 6–6   Variation of MPV with battery temperature.

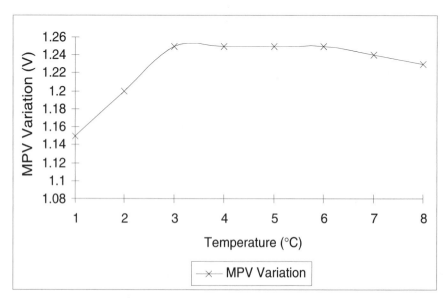

Figure 6–7 Variation of battery cell discharge voltage with discharge capacity.

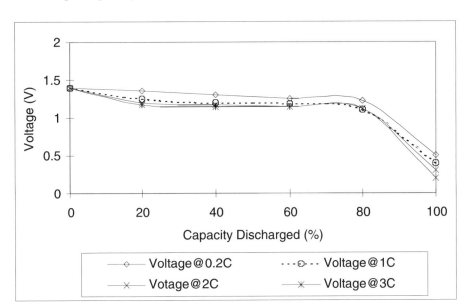

DISCHARGE CAPACITY BEHAVIOR

The available discharge capacity of an NiMH battery is affected significantly both by the rate of discharge and the cell temperature. In addition, the number of formation cycles and the operating history of the battery (i.e., the recent charge, discharge, and storage history of the cell). The useful NiMH battery discharge capacity is the difference between the capacity of the battery restored during the previous charge cycle and the capacity lost due to self-discharge.

Effect of Temperature

As mentioned earlier, the available discharge capacity of a NiMH battery is affected significantly under low cell temperature discharge conditions. As shown in Figure 6–8, the battery capacity increases by almost 50% under the same discharge rate for an increase of 10°C temperature until the battery reaches 20 to 25°C.

During a driving cycle, typically characterized by the Simplified Federal Urban Driving Schedule (SFUDS) discharge, the EV undergoes

Figure 6–8   Variation of battery discharge capacity with temperature.

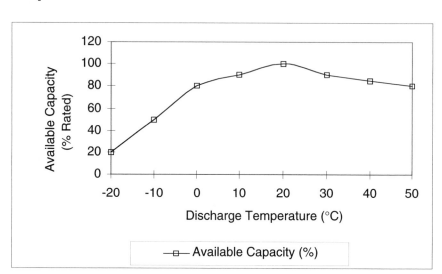

acceleration, constant speed driving, and recovering charge patterns. Under the combination of different conditions, the internal temperature of the battery rises by 25°C. A stable voltage and minimal temperature rise can be attained up to 80 to 90% of the depth of discharge (DOD) under pulse discharge conditions. For a charging current of 10 A applied for 10.5 hours with five- to seven-second pulses at 0.2C, 1C, 2C and 4C, specific power of 160 W/kg at 80 to 90% DOD is attained. The Specific Discharge Power varies with varying cut-off voltages as illustrated in Figure 6–9.

Termination of Discharge

In order to prevent irreversible damage to the NiMH cell due to cell reversal in the discharge, it is recommended that the load be removed from the cell(s) prior to complete discharge. The typical voltage profile for the NiMH cell carried through the total discharge is a dual plateau voltage profile as shown in Figure 6–10. The first voltage plateau is caused by the discharge at the positive electrode while the second voltage plateau is caused by the residual capacity discharge at the negative electrode. Once both the electrodes are reversed, gassing results in the evolution of hydrogen gas, which results in venting of the cells.

Figure 6–9 Battery specific power characteristics at varying cut-off voltages.

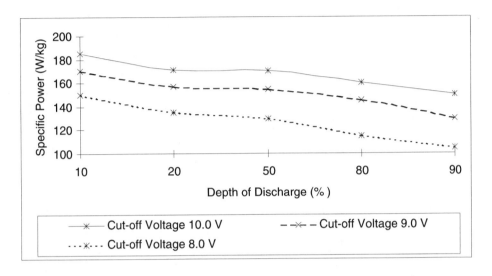

Figure 6–10 Battery cell voltage discharge profile.

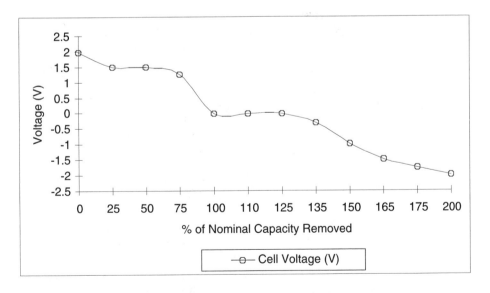

If the evolution of hydrogen gas occurs in excessive quantities, the cells dry out with irreversible structural damage to the electrodes. Thus during discharge, it is important to terminate the discharge at the point where all the cell capacity has been derived based on the first voltage plateau and prior to reaching the second plateau where permanent damage may occur. Since the negative electrode of an NiMH cell absorbs hydrogen, it is less susceptible to long-term damage from cell reversal.

As shown in the Cell Polarity Reversal Profile in Figure 6–10, the true capacity for normal cell discharge is obtained during 0 to 100% removal of nominal battery capacity. After the 100% discharge, the positive electrode undergoes reversal. The second plateau is the start of the overdischarge condition. During the overdischarge from 101 to 200%, the cell undergoes overdischarge leading to reversal of the negative electrode and ultimately reversal of both the electrodes.

Normally the discharge cut-off is based on the voltage drops with a value of 1.5 V per cell (assuming 75% of the 2.0 V per cell nominal midpoint voltage). 1.5 V per cell is an optimal value for most medium to long-term discharge applications (<1C). However, with the high drain-rate usage (at 1 to 4C), the change in the shape of the discharge voltage curve will result in a more rounded "knee" implying that the 1.5 V/cell cutoff may be premature, leaving a significant fraction of the cell capacity unused. Thus a better choice for voltage cut-off under high-discharge rate applications is 75% of the midpoint voltage at that discharge rate. The end-of-discharge voltage (EODV) is dictated solely by considerations of preventing damage to the NiMH cell.

As the cells are combined to form the traction battery, the normal manufacturing results in a range of battery capacities. The effects of the individual cell capacity variations are amplified by the number of cells in the battery. In the series or parallel combination of these batteries, there is a further amplification of the capacity variations based on the number of cells.

Thus the use of 1.5 V/cell × the total # of cells during discharge may result in a weaker cell being driven further into complete reversal before the battery reaches the termination voltage. Both charging techniques that minimize the amount of overcharge applied to the cell and frequent repetitive discharging of the battery may balance the battery pack. Thus minimizing the cell variations leading to premature battery pack failures due to the damage caused by the reversal of the weak cell. The end-of-discharge (or cut-off voltage) provides an acceptable margin to minimize battery failure from repeated cell polarity reversal. Thus, the end-of-discharge voltage (EODV) for a single battery is represented as

$$EODV = [(MPV - 150\,mV) - (n - 1)] - 200\,mV$$

where MPV is the single midpoint voltage at the given discharge rate and n is the number of the cells in the battery. However the safe EODV for a battery pack with total of n_{batt} batteries is the min(EODV) of all the batteries in the pack.

DISCHARGE CHARACTERISTICS OF LI-ION BATTERY

On a module level, the Li-ion battery cells are connected in series and packaged to form modules varying from a 3-cell combination to a 10-cell combination. The module design includes a thermal battery management system based on liquid coolant. This system is able to keep the battery temperature within an optimal temperature range, either by cooling during heavy duty driving conditions or by heating when the battery operating temperature is low.

On a battery pack level, the Li-ion modules are connected in a series combination to form a 300 to 350 V battery pack system for EV application. Each Li-ion battery module is equipped with an electronic module controller. This device combines the function of monitoring the electrical and thermal module data for transmission to the central battery monitor (BMON). The BMON in turn, transmits commands to the battery. Thus the thermal loop and the electronic control devices are both integral to the Li-ion battery. This design of the module improves the safety and the reliability of the battery system.

The high level of battery energy density by weight and volume is suitable for full EV applications. However, the current rate capability of the Li-ion battery system is insufficient for hybrid EV applications. The hybrid EV requires fast discharge and recharge of electric energy. These characteristics are required for acceleration and energy recovery upon regenerative braking. By reducing the electrode thickness, the power capability of the Li-ion cells can be enhanced. The use of thin film electrode for the Li-ion battery design demonstrates continuous discharge characteristics at 15C rate as shown in Figure 6–11.

Under room temperature conditions, a continuous power of more than 850 W/kg at 80% DOD and a voltage level of 85% of the nominal voltage can be achieved. Due to the trade-off between energy and power density, a specific energy of 60 Whr/kg is achieved easily. If the user reduces the amount of inactive components and uses active materials with higher specific capacity, a specific energy of more than 80 Whr/kg can be achieved.

The analysis of the Ragone plot for the Li-ion module, as shown in Figure 6–11 indicates that:

- Prismatic cells exhibit higher specific power and lower specific energy than the cylindrical cells at several discharge current levels.

Figure 6–11   Ragone plot for Li-ion battery.

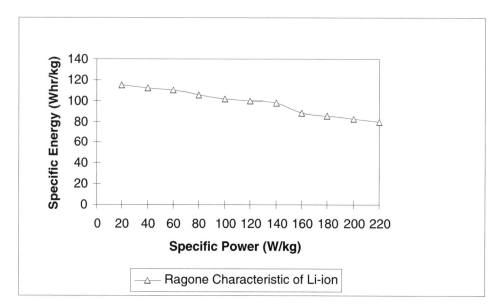

- At higher discharge currents, the cylindrical cells display a better retention of higher energy density than the prismatic cells.
- Both cylindrical and prismatic cells demonstrate low impedance at temperature down to –20°C.

The voltage drop of each cell type is linear with current indicating that the contribution of the interfacial resistance to the total cell imped-ance is negligible. This furthermore indicates that there is a facile charge transfer at the electrode/electrolyte interface as shown in Figure 6–12.

The linear pulse discharge characteristics at different Li-ion module temperatures suggest that the contribution of the interfacial resistance to the internal cell impedance is negligible. The Li-ion cells can be pulsed at very high currents without affecting the cell performance sig-nificantly. These discharge characteristics make the Li-ion battery a suit-able candidate for EV application design.

DISCHARGE OF AN ELECTRIC VEHICLE BATTERY PACK

To illustrate the combined effect of each of the factors affecting the dis-charge of a traction battery, let us assume a well-balanced, fully charged

Figure 6–12   Voltage characteristics of Li-ion under pulse discharge.

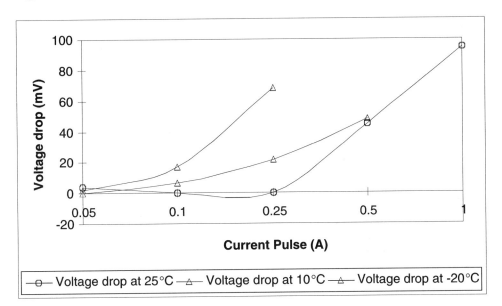

battery pack is being discharged. This means that the driver operates the EV on a highway at 50 mph. Furthermore, assume that the EV with 100 Ahr batteries consumes 1 Ahr/mile or 50 A at a C/2 discharge rate. During the discharge at 50 mph, there is a change in battery temperature that results in the useful or available battery capacity of the pack dropping to 70% of the nominally rated battery capacity. Thus 70 Ahr batteries are now being discharged at C/2 discharge rate, assuming that there is a uniform temperature in the battery pack. With the same consumption rate of 1 Ahr/mile and maintaining the C/2 discharge rate, the battery now consumes 36 A. Owing to further change in the temperature, the new useful driving capacity is 70% of the rated capacity. Thus 70% of 36 Ahr or 25 Ahr is the new useful driving capacity. This gives the driver only 25 miles of driving range before returning to the charging station.

Flooded Pb-acid batteries vent more hydrogen gas throughout their entire lives in comparison with their AGM, VRLA counterparts. Hydrogen is a highly explosive gas. Less than 4% room volume is sufficient to trigger an explosion. This means that the room containing the batteries must have a complete ventilation system with fans and ducts. During normal battery operation, the VRLA battery emits low hydrogen gas levels and does not require special ventilation systems. Although the batteries are operated in enclosed cabinets, without fans or a heating system, the cabinets must be vented or the small emissions of hydrogen will build up to dangerous explosive levels.

In addition, space requirements should also be considered during EV battery pack design. Flooded Pb-acid batteries require 32% more space than their equivalent VRLA battery. The additional space requirements are due to rack requirement and the need to provide space to access the battery for maintenance purposes. Thus when it comes to deciding the most suited Pb-acid battery technology for EVs, the VRLA battery provides the greatest benefits to the user.

COLD-WEATHER IMPACT ON ELECTRIC VEHICLE BATTERY DISCHARGE

As seen with temperature characteristics of the traction battery, low temperature limits the battery discharge and useful available capacity. For commercial viability and customer acceptance, EVs need to operate reliably over a wide climatic range. The cold weather deterioration of range is well known, however, it is important to identify and quantify the causes of battery pack degradation. Once the solutions have been identified, it is important to pursue the solutions to eliminate the causes.

The EV performance under cold temperature conditions is analyzed by installing instrumentation on the vehicle to measure the electrical energy entering and leaving the battery pack. Energy consumed is measured for the system controller, climate or HVAC, and the vehicle accessories.

As part of the early phase of the test plan, it is important to develop a test plan and procedure. The necessary hardware required for the test should be installed to observe the battery capacity and SOC characteristics of the traction battery pack. In addition, it is useful to evaluate the new EV lubricant and tires. Understanding of the winter condition HVAC and accessory loads is also useful in determining the EV performance under cold temperature conditions. While gaining an understanding of the HVAC and accessory loads, it is important to evaluate the correction factors developed during the course of the EV analyses.

The cold weather performance tests are performed at 55 mph driving condition on a level concrete road. The driving profiles used during the performance tests include at least four different drivers with no specific instructions. The battery pack is tested several times during the day of the test. The battery pack tests are then repeated two to three times during the week. All tests of the battery pack performance are terminated when the end-of-test criterion has been reached.

Some of the observations of the cold weather performance on a compact size EV using VRLA batteries are:

1. Aerodynamic drag effect makes a significant impact on cold weather performance. The power required is about 10% higher at 20°F than at 70°F. The aerodynamic drag increases owing to higher air density (for a given drag coefficient) as the battery pack temperature increases.

2. Losses associated with EV lubricants are low at low-operating temperatures. Better performance lubricants are required to reduce the viscous losses.

3. Road traction losses increase under cold weather operating conditions. Newly developed EV battery tires, operating at 50 psi, provide good performance and low rolling resistance at warmer operating temperatures. Further development of battery tires for cold-weather performance will reduce rolling resistance losses.

4. On road traction power required at 0°F is 60% higher than at 70°F. The traction power is indexed to power required at ambient temperature of 70°F. The ratio of power at different ambient temperatures is referenced to 70°F to maintain 55 mph EV speed.

5. Wet versus dry traction power of the EV increases to more than 5% on the wet road. Power required on a wet road to maintain EV speed at 55 mph is approximately 60 mph.

Varying driver profiles impact the EV cold-weather performance as each driver has a different profile. Some drivers are interested in comfort while some of the drivers are interested in distance and performance. The heater accessory power consumption is 1,800 W and the low beam headlights and taillights power consumption is 300 W.

Assuming that the 10 kWhr energy is supplied during the EV driving, at an ambient temperature of 70°F, the compact EV travels 75 miles at 55 mph. Under the same load conditions, the EV travels 50 miles at 55 mph at 20°F. Thus the range of the EV is reduced at a lower temperature.

The battery pack differential—temperature difference—between the ambient and the battery pack is large. In most cases the temperature differential is 15 to 40°C above the ambient temperature. The pack temperature differential is observed to be larger at lower ambient temperatures than at higher ambient temperatures.

The effective battery pack energy diminishes significantly when the Pack Capacity (Whr) tests are conducted at cold temperatures. This is owing to the battery pack temperature and the reduced battery capacity resulting from a higher current draw at higher power requirements. It is important to develop a battery thermal control system to maintain the battery pack temperature within limits.

EV driving range is impacted by temperature. At 55 mph speeds, the vehicle range reduces from 100 miles at 70°F, to 75 miles at 40°F, to 55 miles at 20°F, down to 44 miles at 0°F. This range reduction is due to cold temperature effects on battery pack performance.

When the HVAC system is in use, and the compact size vehicle is being driven at 55 mph, the range reduces to 40 miles. The battery pack in this mode is being maintained at 70°F.

When the HVAC and the Pb-acid battery is under discharge under cold temperature conditions, the range of the EV is reduced to 35 miles at 20°F and 22 miles at 0°F.

The compact EV has a range of 100 miles at steady 55 mph at an ambient temperature of 70°F. Typically, the vehicle consumes 140 Whr/mile at 70°F for propulsion only. The vehicle consumption increases to to 245 Whr/mile at 0°F (75% more).

Assume a gasoline-powered vehicle gets 30 miles/gallon at 55 mph—that's about 1,100 Whr/mile.

If the compact EV power consumption increases by 105 Whr/mile at 0°F, its efficiency would be reduced only 9% to 27.3 miles/gallon. Engine coolant and exhaust system heat may actually reduce the losses.

The compact EV losses may be broken down into component losses. The drive unit mechanical losses are primarily related to the gear train, while electrical losses are related to the alternating current (AC) motor and power inverter. Battery losses are not included. The other chassis losses are due to drive shaft, residual brake drag, and wheel bearings. HVAC is the heating, ventilation, and air conditioning system using a heat pump.

7 ELECTRIC VEHICLE BATTERY PERFORMANCE

THE BATTERY PERFORMANCE MANAGEMENT SYSTEM

A typical electric vehicle (EV) traction battery system consists of a chain of batteries connected in a series, forming a battery pack with nominal voltages ranging from 72 to 324 V and capable of discharge/charge rates of several hundred amperes.

Owing to the fact that no two batteries in a pack are alike, or even that no two cells in a battery are identical or manufactured exactly the same, their parameters—such as capacity—may vary by a few percent. In the case of a new battery, these factors may not be very noticeable, but as the battery undergoes charge-discharge cycles, later on in the battery life these factors determine the performance of the battery pack. In addition, some cells in the battery undergo a change in their parameters such as open-circuit voltage and internal resistance rather abruptly, due to internal dendritic shorts, corrosion, excessive thermal gradients, or loss of electrolyte due to gassing as in VRLA batteries. Such phenomena can lead to hydrogen gas build-ups and may pose a fire or explosion hazard if not detected and acted upon early. This problem may be easily detected in a battery of up to 6 to 12 cells. A faulty cell can be easily disguised in a large battery pack consisting of tens or hundreds of series-connected cells. A similar problem exists for an excessively overdischarged (reversed) cell. Thus for the safety of the EV, it is essential to monitor the batteries individually and detect faults early.

In an EV, the battery of marginally lower capacity than the rest of the pack is the first battery to acquire and indicate a fully charged status. On the discharge side of the cycle, this battery leads the pack and is the first to experience full discharge and reversal of plates. While this battery may not be weaker in any other sense than that it has a relatively lower capacity, it is now the weak link in the chain. This battery will be the first to undergo repeated overcharge and overdischarge, eventually resulting in the failure of the battery.

For a smart monitoring system capable of managing the batteries individually, detecting and isolating a weaker battery is recommended. The Battery Performance Management System (BPMS) quantifies the potential problems associated with an electric vehicle battery pack. BPMS may point to a simple action such as equalization of the charge for either the NiMH or VRLA battery, or suggest replacing a faulty battery to restore the battery pack's full capacity.

The main components of a BPMS include:

- Precision fuel gauge
- Battery charge balance or battery capacity balance for out-of-step batteries and if possible individual cells
- A reference to a standardized data set as the voltage and temperature cut-off control parameter (particularly for a rapid battery charger)
- A data logger for evaluation and processing of battery performance data
- A supervisory data acquisition and control system for battery pack thermal management

In addition, BPMS also has the capability to govern the charge cycle to suit a weak battery. This results in lower utilization of the full battery system capacity, but extends the life of the weaker battery (batteries) and hence improves the life of the entire battery pack. It will also reduce the risk of sudden mode failure. Furthermore, the decline of the weak battery's capacity may be measured and quantified—and when a certain predetermined point is reached, the deteriorating batteries may be re-placed, and the battery pack will be returned to its full-rated capacity. This method of smart charging eliminates the possibility of damage of the batteries due to excessive overcharging during the normal battery recharge cycle and results in a very long cycle life.

A battery in a battery pack can be reduced to a weak state by excessive discharge rates. These conditions of abuse are characterized by short powerful bursts of charging current at excessive voltages during regenerative (regen) braking. Regen can exceed the absolute maximum charge acceptance ability of the battery if it is not properly managed. This condition exceeds the charge acceptance ability of the battery in the range of 80 to 100% SOC (the charge acceptance ability of the battery in 100% SOC is zero). Under these conditions, the battery becomes a large heat sink.

Thus another function of the BPMS is to monitor the discharge or utilization side of the battery to determine the safe operation of the battery—preventing excessive overcharging and overdischarging to allow a more accurate SOC determination as a fuel gauge. Monitoring of the individual battery allows early diagnostics and the detection of a weak or deteriorating battery before its failure. BPMS allows for charge/discharge control matching to the weak battery, preventing its abuse and extending the life of the entire battery. Other characteristics, such as internal resistance readings and their trends, point to a deteriorating battery or even a problem such as poor or corroded contacts (battery interconnects). When predetermined limits are exceeded, warnings can be presented to the driver.

BPMS also extends the concept of a truly smart charging system by placing total control of the battery system on board the EV. BPMS has the capability to both manage the energy flow throughout the operation of the EV, including thermal management of the battery pack, and provide a real-time interface to the power utility infrastructure.

A Model of the BPMS

BPMS, including the defined components, requires a self-adjusting battery model. The algorithm for determining the residual useful battery capacity and the "miles-to-go" includes calculation of the actual battery pack capacity with respect to the nominal battery pack capacity. The Peukert constant is replaced by 10 representative points. Based on the average load current i_{av}, a linear interpolation is applied from the beginning of the battery pack discharge current to the most recent discharge current. In addition, the nominal battery pack current i_{nom} is also monitored and noted.

A correction factor for the capacity at the actual battery temperature $C(T)$ with respect to the capacity of the battery at the reference battery temperature $C(T_0)$ is based on the following equation

$$C(T) = C(T_0) \times (1 + a + b \times \ln(i_{av}/i_{nom}))$$

Calculation of the SOC takes into account losses resulting from the inner resistance R_{int} from the battery voltage V_{batt} under discharge. The no-load battery voltage $V_{no\text{-}load}$ may be expressed as

$$V_{no\text{-}load} = V_{batt} + I \times R_{int}$$

- A second comparison of $V_{no\text{-}load}$ with SOC using a graphical calculation with at least 10 support points for interpolation

- Correction of the actual battery pack capacity as a percentage of the nominal battery pack voltage

The actual battery pack capacity is determined by comparison of the discharged Ahr and the present battery pack capacity with respect to the SOC whenever:

- The battery pack has been fully charged as per the manufacturer's specifications.
- The battery pack discharge capacity exceeds a specified value. Typically 70% of the current rated battery pack capacity.

The Typical BPMS Configuration

From the energy management interface standpoint, the complete system may consist of a pair of main controllers (MC), one serving in active mode while the other is serving as a hot standby. In addition, up to 30 battery monitors (BMONs) interface to the individual batteries, depending upon the battery pack configuration. Each BMON can monitor up to five 12 V batteries in the battery pack and up to four auxiliary temperature and vent pressure sensors. The BMONs must be galvanically isolated from the traction battery pack. The BMONs are connected in a daisy chain, allowing communication via a high-speed data bus (HSDB). A typical system consists of 2 MCs, 7 to 8 BMONs that monitor a battery pack in 12 V increments.

The MC incorporates dual-processor architecture to perform all its functions: communications, processing and control, data storage, and data retrieval. Historical data of the battery pack is stored throughout the life of the battery system, in a nonvolatile RAM. Recent data for the last 30 charge and discharge cycles in full detail forms the first data tier, and older data that is averaged forms the second data tier and is compressed on a biweekly basis or depending upon the usage of the EV. The MC is equipped with data communication interfaces for the charging station and the onboard charger. This allows BPMS to control the fast-mode DC charging, and low rate/overnight charging from a high-voltage AC power grid. In addition to the interfaces, an additional data port provides for system maintenance and remote battery diagnostics.

The individual battery currents are sensed using high resolution Hall effect sensors. These sensors provide for precise summation over time both for the charge and discharge capacity of the battery pack. The MC interface allows for data exchange with the EV controller, and the traction drive train controller via the HSDB interfaces over the EV communication bus. Thus operational limits for the battery current draw, control for regenerative braking as a function of the SOC of the battery are communicated to and from the vehicle bus. The HSDB also provides warnings, alarms, and diagnostic messages to the driver's console.

Auxiliary inputs are interfaced to the MC for direct hardware functions, including the ignition key or park-drive selector interlocks, charging station connector interlock, gas ventilation, and cooling fans. In addition, there is an input for the ambient air temperature sensor. A primary function of the BPMS is to provide system safety. Since BPMS has total control over vehicle charging, all safety features, including driver ignition lock-out, charge-line continuity, charger-polarity check, and line-current leakage are controlled by BPMS prior to applying power to the EV under BPMS control.

BPMS THERMAL MANAGEMENT SYSTEM

The operating conditions of the vehicle batteries consist of variable environmental conditions and variable electric power demands. Chemical processes in the battery are temperature dependent. Therefore, the electrochemical storage system will have to be kept within certain temperature limits in order to maintain a proper function and also to ensure a reasonable battery life. The temperature limits for the battery are considered to be 10°C as the lower limit due to decreasing battery capacity and 50°C as the upper limit due to positive plate corrosion and separator decomposition.

From the battery thermal management standpoint, it is necessary to maintain a uniform temperature within the battery pack. The thermal management system will provide either heating or cooling action depending upon the battery pack conditions. Tests conducted in the laboratory and with EV urban driving suggest that using a thermal management system improves the mileage and battery life by at least 20%. Thermal management of the battery pack is essential both for normal urban driving and rapid overnight charging where charger current levels of hundreds of amperes are applied to the battery pack for relatively short duration.

Although maintenance-free starved electrolyte or gelled-electrolyte VRLA batteries are being used commonly, these batteries overheat more rapidly than the flooded lead-acid counterparts. This is owing to the fact that the VRLA batteries are unable to dissipate heat generated by gas recombination during the charging and I^2R effects during discharging processes.

The primary design of a thermal management system should keep the battery sufficiently insulated. The insulation will assist to obtain an acceptably high-operating temperature during winter and cool during summer by means of an air flow during the charging time. The battery remains in the battery pack during the charging process. The secondary design criterion is that the circulation and the cooling air flow should be properly distributed in space in order to ensure a minimum temperature difference between the individual battery modules. This design criterion should be applicable under various operating conditions.

The thermal capacity of the battery module is derived by calculation and by measurements. The manufacturer-provided information available concerning materials and dimensions is used to calculate the battery pack thermal capacity.

The heat generation of the battery modules is calculated using a simulated driving profile: 185 A in 15 seconds, 61.7 A in 25 seconds, and 0 A in 30 seconds. The charge is performed at 18.5 A until a cell voltage of 2.35 V/cell is reached. After the 2.35 V/cell charge voltage is obtained, the battery is maintained at 9.25 A for 4 hours. The normal discharge lasts for 3 hours, and the normal charge lasts 8 hours plus 4 hours.

The heat generated during the electrical cycling is measured in an isothermal air-flow calorimeter at different temperature levels. The dissipation during discharge is dependent on the temperature level. Starting at approximately 50 W at 50°C decreasing to 25 W at 40°C. The heat dissipation is dependent on the state of discharge with a peak discharge in the beginning, a minimum and a considerable increase in the dissipation at the end of the discharge.

The battery module container analysis accounts for different environmental and operating conditions. The model of the module consists of four parts, a heat transfer model, a heat generation model, an ambient temperature model, and a vehicle operation model. The air-flow model is made by the subdivision of the battery module air-flow stream into several individual sections—10 different standard sections. Each section is characterized by its geometrical properties and the calculations are simplified in order to obtain values for the kinetic and the viscous contribution to the air-flow resistance. The viscous forces are approximated considering a laminar battery air flow in a channel of two parallel plates. The kinetic forces during the air flow are approximated as fully lost by all discontinuities in the battery pack. Each battery flow section is characterized by two constants, k_1 and k_2, expressing the individual section pressure difference Δp as a function of the air stream q. The air-flow difference is expressed as

$$\Delta p = k_2 \times q^2 + k_1 \times q$$

The steady state solution of all the sections is calculated by a computer, using a computer analyses program.

The heat transfer model takes into account the internal heat transport. This heat transfer is due to conduction of heat from the section surfaces through the air over the half air channel width to the air stream. It also accounts for the movement of the heated air by the air stream expressed as a fluid conductance, F as a function of q, the air density ρ the air specific thermal capacity c_p.

The fluid conductance and the radiation between the batter modules is referred to as $F = q \times \rho \times c_p$. The heat transfer also contains conduction through the container walls, free convection from the outer surfaces, and the radiation to the battery pack surroundings. The heat dissipation and the heat capacity of the battery module are also included.

The heat generation model for the battery module takes into account the battery temperature and the charge/discharge status. The relationship is selected as a best-fit linear approximation based on results obtained from the isothermal air-flow calorimeter measurements.

The ambient temperature model describes the ambient temperature as a function of the time of the day. As well as constant temperature as different day/night temperature profiles corresponding to the different seasons.

The vehicle operational model describes the use of the vehicle during the day. It also considers two drive periods—one in the morning and one in the afternoon—and an overnight charge period. Two different modes of use have been considered, one implying the full driving range of the vehicle and one of half the range.

The combined thermal model of the battery pack is used to calculate the temperature response of the battery in the time domain under various conditions. The results were analyzed by considering temperature levels and temperature differences.

Two commonly used thermal management systems are the circulating-air and the circulating-liquid systems. Using either method of thermal management, it is important to maximize performance, minimize cost, complexity, weight, and power requirements. The circulating-air management system can be constructed rather inexpensively around the existing battery pack. As a variation to the circulating-air system, the battery pack may be divided into several sections with individual fans circulating the air in the battery pack. This variation allows for maximum surface exposure and allows adequate air flow around, over, under each battery in the pack. This system adds very little additional weight and requires additional power for the fan and the heater. However, it may be difficult to maintain the temperature gradients to and from the air and the battery pack.

A circulating-liquid thermal management system, provides better cooling and heating than the circulating air system. A liquid solution of ethylene glycol and water as a coolant provides for good heat exchange and reduced volume requirements. The system will become more complex as the heat transfer rates increase. The system is heavier, complex, and expensive compared to the circulating air thermal management system. In addition to the additional weight, the power requirements are more due to additional components like pump and filter.

A complex circulating-liquid thermal management system may include water and ethylene glycol coolant solution jackets between the individual batteries. In addition, it may be necessary to provide for monitoring and thermal management of individual battery cells.

Design Analysis of the Battery Thermal Management System

Air flow at 30 liter/sec should be maintained among the battery pack modules if the battery pack has to be independent of the low winter temperature. In order to attain the air flow of 30 liter/sec, it is necessary to make additional design changes to improve the air flow.

The position of the exhaust opening or openings is not fixed. The analyses shows that concentrating the outlet as far as possible from the intake and placing it at the bottom of the battery pack will provide the most uniform cooling without implying complicated air-guiding arrangements.

The distribution of air between the different battery modules is critically dependent on the small gaps along the sidewalls of the battery modules. In order to compensate the battery module dimension tolerances, a flexible tube is introduced as a separation rod between the two rows of the battery modules. This arrangement takes into account the ribbed design of the battery module, ensuring good mechanical contact and leaving a well-defined flow space in between.

The air tightness of the prototype battery module cover is important to ensure that the overpressure battery ventilation does not expose the sulphuric acid fumes in case of a battery leak.

The battery module design can continue to undergo improvements. The temperature level could be increased by the user adding insulation to the outer walls of the battery module. The cooling period could be extended to 12 hours, making a colder climate acceptable without indoor garage and charging facilities.

The optimum battery temperature for the vehicle application depends upon the separator decomposition and positive plate corrosion. Both the decomposition and the plate corrosion increase by increasing temperature—50°C is considered a suitable upper limit. In addition, the battery cycle life tends to increase by increasing temperature due to increased capacity and thus decreased DOD for the same utilized capacity. In an EV application, the user can expect this effect to be rather pronounced due to the high peak discharge currents.

THE BPMS CHARGING CONTROL

Specialized integrated circuits, available today, have been designed for developing a control scheme for optimization of battery charging. The circuits operate a general assumption that the battery cells share uniform charge and discharge characteristics thus limiting the treatment of the battery as a two-terminal energy storage element.

As discussed in the earlier chapters, limitations in the cell manufacturing process result in no two cells being identical, which leads to uncertainties in the cell characteristics. Furthermore, two detrimental effects of this nonuniformity are that certain battery cells undergo overcharging while the useful charging capacity of the battery decreases. It is essential to minimize the effects of destructive overcharging while maintaining a uniform charge across a battery regardless of the initial cell conditions. A technique, referred to as active equalization allows for a portion of the charging current to be diverted past certain battery cells so that the cells can receive the charging current selectively.

Commonly, DC-to-DC converters are used to shunt current around cells (or a group of cells) in a battery. As the string of batteries charges, each cell in the battery reaches a threshold voltage. Upon reaching the threshold voltage, 15.5 V typical for a 90 Ah VRLA battery, charging current is diverted around the battery. Thus the fully charged battery maintains a threshold terminal voltage, and the excess energy is placed back into the charging bus and appears as additional charging current. The process of recirculating the charging current via shunts allows the undercharged batteries to gain the equalization charge while the fully charged batteries are not overcharged, which in turn prevents gassing and loss of water.

BPMS Charge Protector

In order to ensure maximum life of the battery pack, end-point reliability, and driver safety, VRLA and NiMH battery chemistries require that they be charged and discharged within defined limits. The user can prevent overcharging, undercharging, and discharging by using protection circuits.

The first level of battery pack protection is typically provided using an integrated circuit and a series of Field Emission Transistors (FETs). The battery pack voltage and the discharge are closely monitored (at a cell level if necessary). The battery pack is disconnected from the charger in case the voltage or the current falls outside the specified limits. Typically, the primary electronic circuit does not detect every potential fault. Most silicon-based devices do not detect an overcharging current because it is always smaller than the overdischarge current. Instead, the battery overvoltage is monitored. In most cases, the battery overcurrent sensing circuit does not activate as the charger electronics prohibits further charge after an overcharge. This overcharged battery pack condition is the point when a short circuit is most damaging.

In addition, the charger electronics do not monitor the battery pack temperature at a cell level. The battery pack temperature is monitored by including a Negative Temperature Coefficient (NTC) thermistor. It is under these conditions that a passive Polymer Positive Temperature Coefficient (PPTC) thermistor is included for additional protection. In the event of a failure of the battery pack electronics, the PPTC device limits the cell charge or discharge current within specified safe levels. The secondary PPTC overcurrent protection deadband is usually set above the primary electronics limits to ensure that the circuit is active only as a backup and to prevent nuisance tripping of the battery pack charger.

The PPTC thermistor protects the battery pack charging circuit by rapidly going from a low-resistance state to a high-resistance state in the event that there is an overcurrrent beyond the specified safe temperature limits as shown in Figure 7–1. The PPTC resets itself once the power to the circuit is removed.

Protecting the Traction Battery Pack

In order to improve the sensitivity of the PPTC to higher current densities, to over-temperature conditions during charge a new low temper-

Figure 7–1   PPTC thermistor resistance profile.

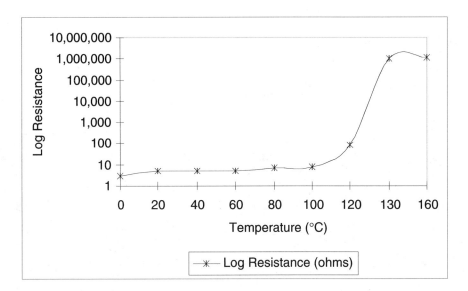

ature PPTC thermistor has been developed. The new low temperature thermistor is now capable of cutting off charging currents to the cells in case the battery pack temperature increases to unsafe limits. This condition may be owing to abusive charging or due to a cell failure in the battery pack. The new PPTC thermistor offers a variety of thermal cut-off resistance values, depending upon the temperature, for various charge currents as shown in Figure 7–2.

The traction batteries can withstand an internal temperature as high as 70°C to 80°C. However, the cell temperatures should be limited to below 100°C and maintained close to the manufacturer's specifications.

In order to prevent the battery cell damage due to thermal runaway, it is necessary to terminate the charging current prior to a rapid cell temperature rise. A PPTC thermistor is placed in series between the charger and the NiMH battery cell. The PPTC thermistor is welded to the negative end of the cell casing. The body of the PPTC thermistor is bonded to the casing to ensure that the heat conducted by the casing is transmitted to the PPTC. The PPTC thermistor welded to the battery cell casing trips at a temperature of 75°C thereby limiting the charging current to below 200 mA.

The cell voltage rises to 4.2 V rapidly, and the temperature of the cell is stable at 30°C. After 20 minutes, the cell voltage rises further to 5.0 V

Figure 7–2   PPTC thermistor profile at varying current densities.

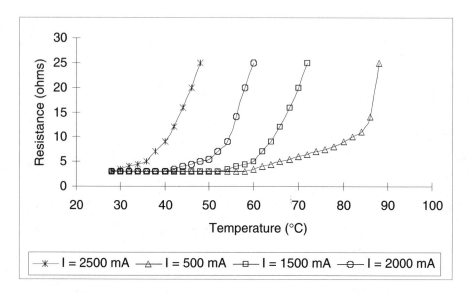

with insignificant rise of battery cell temperature. After 40 minutes, the battery temperature rises at a rate of 11°C/minute. After 44 minutes, the battery voltage and temperature rises. The temperature increase is at a rate of 20°C/minute. The external cell temperature reaches 85°C before stabilizing at 80°C. The maximum cell temperature of 85°C is 30°C below the temperature reached by the battery cell without a PPTC thermistor protection.

The PPTC thermistor operating at a low temperature limits the charge current close to the functional pack operating temperature as shown in Figure 7–3. The PPTC thermistor resets itself when the battery pack temperature rises, owing to excessive sunlight exposure. This feature prevents the battery pack from being disconnected owing to nuisance tripping at high battery pack temperatures.

The BPMS Charge Indicator

The battery charge indicator or fuel gauge should provide the actual battery capacity and nominal battery capacity readings. This indication is represented as:

- A miles-to-go indicator or a fuel gauge
- An economy range indicator in terms of kWhr/mile or kWhr/km

Figure 7–3   PPTC thermistor voltage/temperature characteristics.

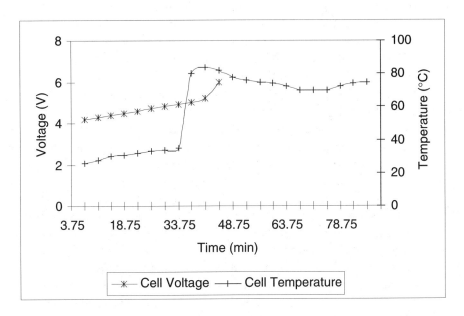

- A warning light or an audible signal for a battery in a dangerous or faulty condition requiring immediate servicing as a "maintenance required" command

This condition should not be capable of being bypassed without a reset and disengagement of the battery pack from the traction controller module. The available energy or capacity of fully formed traction battery can be divided into three portions. The first portion of the capacity is the energy that can be restored or replenished by charging. The second portion of the traction battery energy is the available energy under the present conditions of SOC, discharge, and temperature. The third portion of the traction battery energy is the unusable energy owing to crystalline oxide formation, also known as memory. Both VRLA and NiMH batteries exhibit this memory effect.

While the SOC indicator or fuel gauge is useful, the gauge is reset to 100% each time the battery pack is recharged. The gauge shows a 100% each time regardless of the individual battery's state of health. This leads to a serious miscount of the battery pack energy being shown as 100% after a full charge, when in fact the charge acceptance has dropped down to 70 to 50%. Thus it is important to understand the indicator of battery pack capacity when it indicates 100%. The question that arises is "100% of what battery capacity?" A user unfamiliar with the battery pack capacity may find that the EV fuel gauge or miles-to-go indicator may show a 100% after the battery pack undergoes an overnight charge. However, when the fully charged EV has been driven under cold climatic conditions, for a few miles and stopped, the fuel gauge shows only 80 to 75% upon restarting the vehicle. Thus the only practical way to measure the state-of-health is by counting the energy units (coulombs) of the battery pack. A perfect battery that is fully charged will indicate 100% on a calibrated fuel gauge. The state-of-health of the battery pack is determined by the state of the weakest battery in the pack. It is the discrepancy between the factory-set 100% and the actual delivered coulombs that is used to calculate the state-of-health of the battery pack.

The state-of-health and the SOC of the battery pack can be used to calibrate a linear battery fuel gauge. The first portion of the battery pack representing the actual available energy is represented by green LEDs. The second portion of the battery pack representing the empty or rechargeable portion of the battery is represented by a set of dark LEDs. The third portion of the battery pack representing the nonrechargeable portion or unusable battery capacity is represented by red LEDs.

Smart battery chargers can check the state-of-health of the battery pack. The charger monitors the previously delivered power of the battery pack and compares it with the target capacity of the battery pack. The target capacity of the battery pack can be regulated especially during the early formation cycles of a relatively new battery pack. Adjustable to 60%, 70%, or 80%, the target capacity acts as a battery pass/fail criterion. Based on the target capacity threshold, the weaker battery below the setting is flagged. An additional "battery condition" button prompts the driver of the EV to recalibrate the battery pack. The recalibration or capacity relearning of the battery pack consists of a charge/discharge/charge cycle of the weaker battery in the pack.

A smart onboard battery charger can be configured to apply the conditioning discharge whenever required. The discharge override button cancels the battery pack discharge in the event a fast-charge is required. In the event the fast-charge condition arises, it indicates that a battery or set of batteries need replacement.

Depolarization as a Process to Enhance Charging

Battery charging relies on the principle of moving ions from a positive plate to a negative plate. It is the efficiency of the ion movement over the cell that impacts charging efficiency. The recombination process within a battery cell can be illustrated as an electrode generates ions and the electrode consumes it. If the electrode that is consuming the ion consumes it at a fast rate, then there are no additional ions available for sustaining the charging process. The electrode is starved of ions and waits until more ions move into its vicinity.

In order to avoid this, the charger delivers a 2.5 A charging pulse to each battery for approximately 500 msec. The BMON controls the charge pulse to individual batteries in the battery pack. As a result of the charging pulse, ions start to build up at a fast rate, limiting the charge acceptance of the battery. In order to avoid the limiting of the charge acceptance, a negative 11 A pulse is applied for a shorter duration (typically 2 msec). The duration and frequency of the pulses are varied to effectively rebalance the cell's ion concentration. This improves charge acceptance and permits a more efficient charging process.

Although a battery is not discharged by a short negative pulse, the application of periodic negative pulses eliminates the need to pre-discharge NiMH batteries. Normally the batteries have to be fully discharged before the application of a charge in order to avoid the dreaded memory effect. This is the condition exhibited by both the VRLA and the NiMH batteries in any state of charge.

During this condition, the negative electrode changes its metallurgical state from what is called alpha to beta, and it is no longer capable of delivering power. The battery exhibits a memory of its last charge capacity.

Discharging the battery to an abnormally low level converts nickel back into an alpha state that can be recharged. With the depolarizing pulse, the cell voltage drops to a level that converts the nickel plate material back to original form ready for normal operation.

As the battery charging progresses, the BMON monitors the battery current, voltage, and temperature. The BMON, which includes the analog to digital converters and the switched mode power supply, determines the battery pack's ion concentration at the individual battery level during the charging and discharging process. In addition, the BMON varies the pulse frequency based on the ion concentration estimates.

Smart Battery and BPMS Diagnostics Control

A smart battery removes the charge control from the battery charger and assigns it to the battery. With the intelligent high speed data (HSD) bus, the battery becomes the master, and the charger behaves as the slave that follows the battery. This approach eliminates the recalibration and configuration of the charger with new charging data profiles based on manufacturer, new/old battery chemistry, and compatibility.

Charging of a battery with minimal built-in intelligence is quite analogous to an infant being fed by its mother. The reaction of the infant during feeding determines whether or not the mother provides additional food. On the contrary, the intelligent battery can inform the charger as to how much additional charge or discharge is required to maintain optimal performance of the traction battery. The battery diagnostics information is transmitted using the smart HSD bus. The diagnostics battery pack information sent to the BMONs includes battery type, serial number, manufacturer's name, and manufacturer's date.

In addition, EV service stations can download battery pack charging data. This data is useful to locate the fault and initiate a corrective action for:

- Conditioning or repair of battery modules
- Replacement of faulty battery components, including connectors, fuses, and wiring
- Replacement of an entire battery module

In addition, the intelligent battery pack information sent to the BMONs includes battery cycle count, vehicle driving profile, or driving pattern.

HIGH-VOLTAGE CABLING AND DISCONNECTS

EVs have two different wiring systems: high-voltage and low-voltage. The high-voltage wiring system is used primarily to provide energy to the motor to propel the vehicle. However, some vehicle manufacturers use high voltages to power heating/cooling systems, power steering pumps, and EV sensors. The Society of Automotive Engineers (SAE) has specified that orange cables are the standard color for high-voltage wiring in EVs.

At present, EVs have high-voltage systems ranging from 250 to 360 V DC and in some cases even higher. The high traction voltages are provided by battery packs composed of dozens of individual batteries wired in series and sealed in protective cases.

A separate 12 V auxiliary battery is part of design. This auxiliary battery is used for accessories such as vehicle instrumentation, lighting, and HVAC. A separate 12 V battery is kept charged by a DC-to-DC converter that steps down the voltage from the high-voltage traction batteries.

In EV designs, major automobile manufacturers use isolated electric busses for both the positive and negative sides of the high-voltage electrical system. This is an important safety feature. In the event of positive electric bus isolation loss with respect to the vehicle frame or chassi, no electrical current passes through the vehicle frame or chassis. As a result, vehicle drivers or emergency responders will not be subject to a hazardous shock by the accidental loss of isolation between the positive or negative electric busses with respect to the vehicle frame or chassis. The EV system design differs from internal combustion EV systems because the 12 V DC system relies on the vehicle frame and chassis as the negative electric buss. This is an acceptable wiring design because people are not exposed to lethal voltages or currents with 12 V DC systems.

In addition, all OEM EVs have automatic high-voltage system disconnects as a primary safety design feature. These automatic disconnects include a combination of ground fault monitoring, an inertia switch, and/or a pilot circuit.

Ground fault monitoring disconnects operate on the same principles as the ground fault monitoring devices used in everyday households circuits. These devices monitor the ground system in the EV for current leakage from the high-voltage battery pack. If a fault in terms of current leakage is detected, the devices automatically disconnect the high-voltage system from the battery system. The location of the ground fault monitoring system varies with each vehicle design, but it is typically in the vicinity of the traction battery pack.

In the case of EVs that use the inertia switch disconnect, the end result is the same but the method of isolation is slightly different. The inertia switch senses high-deceleration rates such as those encountered in a vehicle accident. In the event rapid deceleration occurs, the inertia switch is automatically tripped and the high-voltage system battery pack is disconnected from the rest of the EV. The inertia switch is set for a low impact. Inertia switches are also common on internal combustion engines and are used to de-energize electric fuel pumps from the engine in the event of an accident. In most cases, the inertia switch can be reset by pressing a button located on the device. Although the location of these switches varies based on vehicle design, most are located in the motor compartment.

Other vehicles use pilot circuit disconnects; again, the end result is the same, but the mechanism is entirely different. Throughout the motor compartment, high-voltage cables are routed between the battery pack, the electronic control module, the motor, the battery charging port, and other high-voltage components. Running parallel to these high-voltage cables is an additional pilot circuit that acts as a simple continuity loop. The pilot circuit is integral to the high-voltage charging cable such that it is not possible to disconnect. In the event that a break occurs in the charge cable it happens by doing the same to the pilot cable. If an accident occurs that results in the high-voltage and causes the pilot cable to disconnect, the pilot circuit records the loss of electrical continuity. It will automatically disconnect the high-voltage cabling from the battery pack. The location of this pilot circuit disconnect system is also vehicle-specific but is typically found in proximity to the vehicle battery pack.

Many vehicle manufacturers employ a combination of the disconnect systems for both redundancy and safety purposes. Whether a ground monitor, inertia switch, or pilot circuit is used, it is important to know that these devices isolate the rest of the vehicle only from the traction battery pack voltage. However, lethal levels of electric current may still be present in the battery pack. It is of utmost importance that an EV battery pack be treated with the same caution and respect as a full gasoline fuel tank in an internal combustion vehicle.

All current OEM EVs also have special manual disconnects that decouple the battery pack from the remainder of the vehicle wiring and systems. The locations of these disconnects are very vehicle-specific and are intended to be used mostly by vehicle service personnel during periodic maintenance procedures. Newer electric bus models now have a manual disconnect located on the driver's control panel. This allows for additional safety in the event that the inertia switch, ground monitor, or pilot circuit fail to disconnect the battery pack from the vehicle wiring systems.

SAFETY IN BATTERY DESIGN

Battery electrolyte decomposition can be hazardous to the EV operator. Overheating of the traction battery pack accelerates the electrochemical reaction that causes electrolyte decomposition. During the first charging cycle, the process of initial formation of the interfacial films leads to the electrolyte reduction. This reduction may continue in to the subsequent charging cycles with certain combinations of the negative electrode and the electrolyte materials.

In addition, electrolyte decomposition leads to phase changes, which can also pose hazards to the EV operator. The organic liquids identified for the possible use of the VRLA and NiMH battery electrolytes have boiling points in the range of 60 to 250°C under standard conditions. The boiling point of any proposed electrolyte is a key factor, as vaporizing the electrolyte will damage the cell integrity.

For example, vehicle passengers can be exposed if battery containment fails and the electrolyte leaks. Battery pack electrolytes are more likely to leak when a new cell is damaged. New cells contain more electrolyte than previously used cells, because some electrolyte is consumed during cycling. Exposure can also occur during processing of used traction batteries. Overheating, overcharging, and overdischarging can cause decomposition or phase changes in the electrolyte, posing hazards of exposure to the electrolyte decomposition or gaseous electrolyte compounds.

Exposure to other cell materials can also occur during the manufacturing of the batteries. Once battery cells are completed, exposure to aluminum, copper, and nickel will be unlikely. Battery overcharging and venting can cause exposure to the fumes from the decomposition of polypropylene or polyvinylidene fluoride.

Acid spills from the battery pack are also an important factor for battery pack design. A typical flooded Pb-acid battery has 15 to 19 gallons of corrosive electrolyte. A set of 24 batteries contains 360 gallons or 3,600 pounds. Since the electrolyte is corrosive, UBC 307.2.3, The Uniform Fire Code, Article 64, and the local codes that reference the fire code require that the entire battery be surrounded with an acid containment system, especially when the battery exceeds the acid volume limit of more than 100 gallons.

This system adds another $1,700 to $3,000 for each rack of the flooded Pb-acid batteries. In comparison, the VRLA battery has no free electrolyte and thus requires no additional systems. This further translates to additional savings of $1,700 to $3,000.

The VRLA battery pack design has a higher initial cost. However, when the cost of maintenance, ventilation, installation, etc., is factored into the overall cost of battery pack, the VRLA based battery pack is 21 to 36% cheaper than the flooded Pb-acid battery during its entire life. The worst-case scenario of 21% is based on replacing the VRLA battery after 15 years in comparison with 20 years with the flooded Pb-acid battery.

Electrical Safety

Using electrical safety as an example, the EV connector must be polarized and configured so that it is noninterchangeable with other electrical devices such as electric dryers. The method by which the EV charging equipment couples to the EV can be either conductive or inductive, but must be designed so as to prevent against unintentional disconnection. Additionally, the new electrical codes require that EV charging loads be considered continuous; therefore, the premises wiring for the EV charging equipment must be rated at 125% of the charging equipment's maximum load.

All EV charging equipment must have ground-fault circuit interrupter devices for personnel protection. Rainproof the battery system, including the battery pack, for outdoor compatible equipment. An interlock to de-energize the equipment in the event of connector or cable damage must be incorporated. Furthermore, a connection interlock is required to ensure that there is a nonenergized interface between the EV charging equipment and the EV until the connector has been fastened to the vehicle.

A ventilation interlock is also required in the EV charging equipment; this interlock enables the EV charging equipment to determine whether a vehicle requires ventilation and whether ventilation is available. If ventilation is included in the system, the ventilation interlock will allow any vehicle to charge. However, if ventilation is not included in the system, the mechanical ventilation interlock will allow vehicles equipped with nongassing batteries to charge, but not vehicles equipped with gassing batteries.

Mitigation of Intrinsic Materials Hazards

The intrinsic material hazards of some of the traction battery designs can be mitigated through battery design and workplace procedures.

Using integrated circuits to monitor battery cells may assist with, both electrical and thermal management. In case of the battery pack, the individual battery temperature and current is monitored by using battery monitoring systems (BMONs).

As mentioned earlier, intrinsic material hazards increase when VRLA and NiMH batteries are exposed to elevated battery pack temperatures. This can cause hazardous conditions such as exothermic and gas-producing reactions. In an EV, heat from the battery itself, and other components can lead to elevated battery pack temperatures. The thermal management system mitigates the hazards caused by elevated battery pack temperatures. Research trends suggest that thermal runaway with heat sensitive (shut-down) separators will be able to stop electrochemical reactions.

The electrical system abuse poses material hazards. Short circuit of cells raises the battery temperature. The battery temperature rise accelerates the reactions between the negative electrodes and the battery electrolyte. Select abuse-tolerant materials and protecting cells within the battery against overcharge and overdischarge to mitigate the hazards of exposing battery materials to high temperatures.

Overcharge and overdischarge protection may be achieved through adjusting the battery cell chemistry to minimize the effects of overcharge and overdischarge using protective battery electronics. In addition, the battery cell chemistry may be adjusted to protect the electrolyte from cell oxidation during battery overcharge. By introducing an electrolyte additive, the reaction will reversibly oxidize above the normal maximum positive electrode potential and below the potential at which the bulk electrolyte material oxidizes. Overcharge and overdischarge protection may be improved by ensuring that each cell contains a chemically balanced amount of positive and negative electrode materials. Battery protection electronics also provide cell protection against battery overcharge and overdischarge by using a combination of battery pack fuses and internal safety mechanisms. Using smart battery protection, the electrical safety mechanism operates during accidental overcharge, when gas evolves. Although normal cell reactions do not generate excessive amounts of gas, pressure build-up causes venting due to gas formation. This in turn leads to failure of the traction battery due to loss of electrolyte. The mechanism operates upon raising internal cell pressure until a vent opens. This in turn breaks the battery circuit.

In a battery pack, an individual battery using an organic electrolyte may cause hazardous electrolyte spills in the event the cells are damaged. The design of the cell and battery container requires seals that can prevent electrolyte spills. Optimizing the amount of battery electrolyte can limit the severity of spills that occur in addition to battery seals. The amount of electrolyte required to conduct ions throughout the life of the cell must also account for the electrolyte decomposition during cycling.

BATTERY PACK SAFETY—ELECTROLYTE SPILLAGE AND ELECTRIC SHOCK

The EVs currently produced worldwide carry a large number of traction batteries onboard. Therefore, a large amount of electrolyte is in either liquid or gel form. In the event of an EV accident, a rollover or crash, there is an associated hazard associated with exposure to such a large amount of electrolyte. This hazard further extends to vehicle occupants, neighboring vehicles, bystanders, and emergency and clean-up personnel. Some of the important issues that must be addressed in understanding what types of traction batteries are expected to be in production use over the next 5 to 10 years, including their form (liquid or gel type electrolyte), chemical properties of the traction batteries, and associated battery pack temperatures of the various electrolyte solutions are:

- What is the nature of the electrolyte solutions in terms of their pH—namely are they acidic, alkaline, or water reactive solutions?
- Where are the battery packs located in the EV?
- What are the safety problems associated with the electrolyte contact in the event of a rollover spillage to EV occupants, rescue teams, or clean-up personnel?
- Can battery electrolyte spillage result in potential fire hazard or thermal electrolyte burns?
- Can the battery electrolyte spillage result in toxic or asphyxiant vapors?
- Under what conditions can an electrolyte spillage serve as an electrical conductor or short circuit, thereby creating a fire hazard?
- What are the potential safety consequences of having spilled electrolyte from an EV crash mix with a different electrolyte or vehicle fluid including gasoline, diesel, engine coolant, or oil?

Furthermore it is important to:

- Determine the amount of electrolyte spillage allowed after a crash or rollover.
- Determine the requirements for the spillage of high temperature liquid coolants from the EV batteries.
- Determine what locations of the traction battery pack minimize the battery electrolyte spillage.
- Determine if the traction battery pack should use a dual-walled design such that in the event of a rollover, damage of the outer wall of the battery pack will not result in electrolyte spillage.
- Determine if there should be sufficient labeling inside the battery pack—the EV—to better assist emergency rescue teams at the scene of the EV crash.
- Determine the electric shock hazards associated with an EV. Since most EV powertrain systems operate under relatively high levels of electric power, 600 V, 550 A maximum. There is a potential for electric shock to persons associated with EV repair and maintenance personnel.

CHARGING TECHNOLOGY

With EVs comes the EV recharging infrastructure, both for public, domestic, and private use. This charging infrastructure includes recharging units, ventilation, and electrical safety features for indoor and outdoor charging stations. To ensure the safe installation of charging equipment, changes have been made to building and electrical codes.

Charging Stations

During EV charging, the charger transforms electricity from the utility into energy compatible with the vehicle's battery pack. According to Society of Automotive Engineers (SAE), the full EV charging system consists of the equipment required to condition and transfer energy from the constant-frequency, constant-voltage supply network to direct current. For the purpose of charging the battery and/or operating the vehicles electrical systems, vehicle interior preconditioning, battery thermal management, onboard vehicle computer, the charger communicates with the BMON. The BMON dictates how much voltage and current can be delivered by the building wiring system to the EV battery system.

Charging of the battery pack is passing an electrical current through the battery to reform its active materials to their high-energy charge state. The charging process is a reverse of the discharging process, in that current is forced to flow back through the battery, driving the chemical reaction in the opposite direction. The algorithm by which this is accomplished is different for each battery type due to the variations in the batteries' chemical components.

The EV is connected to the Electric Vehicle Supply Equipment (EVSE), which in turn is connected to the building wiring. The National Electrical Code (NEC) defines this equipment as the conductors, including the ungrounded and grounded, equipment grounding conductors, the EV connectors, attachment plugs of all other fittings, devices, power outlets, or apparatus installed specifically for the purpose of delivering energy from the premise wiring to the EV.

For residential and most public charging locations, there are two power levels that will be used: Level 1 and Level II. Level I, or convenience charging, occurs while the vehicle is connected to a 120V, 15A branch circuit, with a complete charging cycle taking anywhere from 10 to 15 hours. This type of charging system uses the common grounded electrical outlets and is most often used when Level II charging is unavailable. Level II charging takes place while the vehicle is connected to a 240 V, 40 A circuit that is dedicated for EV usage only. At this voltage and current level, a full charge takes from 3 to 8 hours depending on battery type. EVSE for this power level must be hardwired to the premises wiring.

A third power level, Level III, is any EVSE with a power rating greater than Level II. Most of the Level III charging system is located off the vehicle platform. During Level III charging, which is the EV equivalent of a commercial gasoline service station, an EV can be charged in a matter of minutes. To accomplish Level III charging, it is likely that this equipment may be rated at power levels from 75 to 150 kW, requiring that the supply circuit to the equipment be rated at 480 V, 3 φ, 90 to 250 A. Supply circuits may require to be even be larger. Only trained personnel should handle this equipment.

All EVSE equipment, at all power levels, are required to be manufactured and installed in accordance with published standards documents such as: NFPA (NEC Article 625), SAE (J1772, J1773, J2293, others), UL (2202, 2231, 2251, others), IEEE / IEC, FCC (Title 47–Part 15), and several others.

Coupling Types

EVSE can be connected to the EV by the general public under all weather conditions. There are currently two primary methods of transferring power to EVs: (1) conductive coupling, and (2) inductive coupling.

In the conductive coupling method, connectors use a physical metallic contact to pass electrical energy when they are joined together. Specific EV coupling systems—connectors paired with electrical inlets— have been designed that provide a nonenergized interface to the charger operator. Thus, not only is the voltage prevented from being present before the connection is completed, the metallic contacts are also completely covered and inaccessible to the operator.

In the inductive coupling method, the coupling system acts as a transformer. AC power is transferred magnetically, or induced between a primary winding, on the supply side to a secondary winding on the vehicle side. This method uses EVSE that converts standard power-line frequency (60 Hz) to high frequency (80 to 300 kHz), reducing the size of the transformer equipment. The inductive connection is developed primarily for EV applications, though it has been applied to other small appliances.

In both conductive and inductive coupling, the connection process is safe and convenient for all EV applications.

ELECTRICAL INSULATION BREAKDOWN DETECTION

The breakdown in electrical insulation of the battery pack terminals can lead to a leakage current flow between the high voltage system and the vehicle chassis. A high voltage arc results in a fatal shock. In the event of an insulation breakdown, the detection circuit generates a fault signal trigger to the BMON. This fault signal is generated when the specified threshold of leakage current from high voltage to the battery chassis is detected. Since the high battery pack voltage is floating with respect to the ground, there is no current flowing to the chassis.

The electrical breakdown operates in a voltage range defined by the low-voltage operating voltage, high-voltage operating voltage, break-down survivability low-voltage, and high-voltage.

A control signal line, also referred to as the pilot line, assures that the high voltage connector is plugged in while the vehicle is in operation.

ELECTRICAL VEHICLE COMPONENT TESTS

EV components are evaluated for performance based on the vehicle engineering standards. The components are tested based on the standards including:

- Reliability/durability test
- Life Test
- Extended Life Test
- Mechanical Shock Test
- Mechanical Vibration Test
- Thermal Shock Test
- Humidity Soak Test
- Electromagnetic Compatability Test
- Random Drop Test

Reliability/Durability Test

The reliability and durability test includes recording all incidents related to component testing. Before the testing begins, the test samples are checked and approved for functional requirements. During testing, the test samples must be mounted in the vehicle position. The mounting fixture and test equipment are subject to approval by vehicle engineering.

The test samples must be monitored at least once every hour during normal working hours. All incidents must be recorded.

Operating Life Test

The EV components are exposed to a temperature/humidity cycle for a period of 1,500 hours of total test time. The components are monitored during the test using an appropriate mounting:

- Switch the power on. Stabilize the component temperature at 38°C and maintain the relative humidity at 90% (noncondensing) for one hour.
- Discontinue the addition of moisture and increase the temperature to 85°C over a period of 0.5 hours.
- Stabilize the component temperature at 85°C for 2 hours. Soak the component in power on condition for 1 1/2 hours. Power off the component for 30 minutes. Cycle through the soak test for the component.
- Cool the component down to −40°C over a period of 1 1/2 hours. Humidity in the chamber may condense and form water on the component surface. Protection of the component from condensation may be required to simulate the actual mounting conditions in the vehicle.
- Allow the component to stabilize at −40°C for 1 1/2 hours with the component in the ON condition and continue the soak test for an additional 30 minutes.
- Allow the temperature to drop and rise back to 38°C for 1 1/2 hours. Stabilize the component at 38°C and continue the soak.
- Repeat the test for 1,500 hours of total component test time.
- Allow the component under test to return to the ambient temperature.
- Verify the proper operation and inspect the component for structural damage.

Extended Life Test

The extended life test represents one design life (10 years/100,000 miles). During the operating test, the component is tested for three times its life. Continue the component test five of the test samples to failure or three times life, whichever occurs first.

Vehicle Endurance Test

Components for the EV are tested to endure 50,000 miles of Powertrain endurance or 30,000 miles of the Arizona Vehicle Endurance Test.

BUILDING STANDARDS

To ensure that the charging equipment supporting EVs is safe, the National Electric Vehicle Infrastructure Working Council (IWC) was formed to address EV infrastructure issues. The IWC is a consortium of representatives from across the nation and around the world, representing industries, electric utilities, automotive engineers, electrical manufacturers, code consultants, EV industry organizations, regulatory agencies, and independent testing laboratories, such as the Underwriters Laboratories (UL).

The IWC has recommended a code that addresses the electrical requirements for EV charging equipment. Along with the SAE, the code proposes for inclusion in the 1996 National Electrical Code (NEC).

These codes address several issues associated with EV charging equipment. These issues can be classified primarily, as pertaining to electrical safety devices required in the equipment or the ventilation of the charging system location.

VENTILATION

Part II, Uniform Building Code, of California Title 24 Code of Regulations addresses location and ventilation issues associated with EV charging. These codes address where EV charging equipment can be installed. If a ventilated charging system is to be installed, the codes specify how much mechanical ventilation must be provided to ensure that any hydrogen gassed-off during charging is maintained at a safe level in the charging area.

The ventilation rates specified in the building codes are calculated to comply with the NFPA requirements published in Standard NFPA 69, Explosion Prevention Systems. This standard establishes requirements to ensure safety with flammable mixtures. Section 3-3, Design and Operating Requirements, requires that combustible gas concentrations be restricted to 25% of the Lower Flammability Limits. This design criterion provides a safety margin in atmospheres containing hydrogen. Hydrogen is combustible in air at levels as low as 4% by volume of air. In order for the charging station to not be classified as "hazardous," the hydrogen concentration must not exceed 10,000 ppm, which equates to 1% hydrogen by volume of air.

8 TESTING AND COMPUTER-BASED MODELING OF ELECTRIC VEHICLE BATTERIES

As the electric vehicle (EV) development continues, it is important to simulate and validate the development of advanced batteries for electric and hybrid-EV applications. From an end-to-end system perspective, it is important to validate the performance of the batteries by defining the performance envelope of the battery pack. This includes subjecting the traction battery pack to in-vehicle testing for electric and hybrid-EV applications. Integrated with the computer-based simulation, the performance analysis provides a baseline for the battery pack in real-world driving conditions.

The end-to-end testing approach overcomes some of the significant existing barriers faced during the development of EVs. The new capabilities include:

- Enabling design improvements into the internal characteristics of the battery, active material utilization, electrolyte utilization, inter- and intra-cell temperature distribution, etc.
- Improve and evaluate the advanced traction battery designs and alternatives prior to building a physical vehicle prototype.
- Predict battery SOC, onboard charging profiles, terrain-based vehicle driving profiles.
- Design virtual battery system designs for vehicle integration and simulation.
- Improve system designs of rapid chargers and battery performance management systems.

The integrated testing includes validation of the EV, Pb-acid batteries, NiMH batteries. The tests include a durability course test, impact pendulum test, emission tests, and crash and obstacle tests found on the roads (such as signs and barriers).

The integrated tests also include dynamometer tests including constant force test, constant power discharge test, and even a 1/4 mile drag race profile test. Data analyzed includes power draw, engine torque, speed, and acceleration of the vehicle. The data analyzed is collected and displayed graphically or exported to an Excel file for analyses.

The actual testing takes about one hour for both the vehicle and the battery testing. While the computer simulations required only one minute of the CPU time on a standard PC.

The VRLA 85 Ahr battery undergoing a complex DST cycling has to terminate the test at the 80% depth of discharge (DOD) in about 76 minutes to avoid battery overdischarge and permanent damage. The virtual simulation can continue beyond the 80% DOD value until the battery is fully discharged (approximately in 100 minutes). The computer simulation helps to determine the ultimate battery limit under the DST cycle or in a real driving cycle in a nondestructive cycle. This simulation demonstrates that the end of discharge of this battery is due to acid depletion at the positive plate. The battery underutilizes the active material by as much as 70%. Since it is virtually impossible to experimentally determine the utilization of active materials, computer-based simulation provides the alternative. The low utilization of the active material indicates an opportunity to improve the energy density and reducing the cost of the existing battery technology.

The NiMH battery model can successfully predict electrochemical, gassing behavior, and the thermal behavior. The cell potential, pressure and temperature of the NiMH battery can also be successfully predicted using computer simulation. The coupled thermal and electrochemical modeling provides a heat generation rate model. This model predicts the heat generation rate due to electrochemical reactions and Joule heating. In addition, it is possible to predict the temporal and the spatial variations in the cell temperature of the battery. These temperature variations strongly affect the electrochemical and transport processes owing to temperature-dependent chemical properties. The coupled models are required to capture thermal runaway phenomenon in advanced Pb-acid and NiMH battery systems.

The coupled tests demonstrate that an electrolyte temperature gradient develops along the battery cell during discharge. The gradient is as much as 40°C at 75% DOD and is exhibitd by Pb-acid, NiMH, and Li-ion battery cells. This large temperature gradient results in nonuniform electrode reaction rates along the cell height. As the cell size increases, this gradient effect becomes significant.

TESTING ELECTRIC VEHICLE BATTERIES

The battery is a key element in the acceptance of the EVs. In order to test and develop the battery pack, it is important to determine the effect of battery cycle life versus the peak battery power and battery rest periods, and to determine the impact of the charge method on the battery cycle life. From the battery pack management standpoint, it is important to study the thermal management characteristics and the utilization of active material in the batteries.

The United States Advanced Battery Consortium (USABC) has defined procedures for battery pack testing. These test procedures include battery pack life-cycle testing, destructive safety and/or special testing, destructive abuse testing, and posttest analysis.

A typical battery test flow includes the following test steps (not necessarily in the same order):

- Core battery performance testing
- Performance safety/abuse or life cycle testing
- Life cycle reporting
- Posttest analysis and reporting

Core Battery Performance Tests

The core battery performance tests include the following:

- Constant current discharge
- Peak power
- Constant power
- Variable power discharge
- Federal Urban Driving Schedule (FUDS) regime
- Dynamic stress test (DST) regime
- Special performance
- Partial discharge
- Standloss
- Sustained hill-climb power
- Thermal performance
- Battery vibration
- Charge completion or optimization
- Fast charging

A brief description of each of the above-mentioned tests provides an insight in to the nature of the battery test and how the test improves the development of the EV from the traction battery pack standpoint.

The Constant Current Discharge Test

This test determines the effective battery capacity of the battery pack using repeatable, standard tests. A series of predefined current levels are applied to characterize the effect of discharge rate on the battery pack. The test is terminated when the rated battery capacity or the minimum discharge voltage (as specified by the battery manufacturer) is reached, or whichever occurs first.

Peak Power Test

This test determines the peak (maximum) battery power capability of a battery at various DOD. The value is calculated at 80% DOD and is important since it provides a comparison with the USABC power goal values. This test does not measure the actual peak power available from the battery.

Constant Power Test

This test determines the ability of the battery to provide a sustained discharge over a range of power levels that represent EV applications. The tests perform a sequence of constant power discharge/charge cycles that define the voltage versus power behavior of a battery. The behavior is as a function of DOD.

At each power level, the battery undergoes a discharge to rated capacity or a specified termination point, whichever comes first, a minimum of two times.

Variable Power Discharge Test

This test produces the effects of EV driving behavior (including regenerative braking) on the performance and the life of a battery pack. The variable power discharge profiles specified under this test are based on the auto industry standard Federal Urban Driving Schedule (FUDS), a 1,372-second time-velocity profile originally based on actual driving data. The schedule has been approximated to include the EV application of battery pack charge/discharge cycles.

Federal Urban Driving Schedule Regime (Variable Power Discharge) Test

This represents the best possible simulated condition of the actual power requirements from the EV standpoint. This test provides a demanding profile with respect to the occurrence of high power peaks and the ratio of maximum regenerative charging, to the discharge power applied to the battery pack.

Dynamic Stress Test Regime (Variable Power) Test

This test provides a simplified profile of the Federal Urban Driving Schedule Regime (FUDS)-based power-time test. This specific profile can simulate the dynamic behavior of the EV undergoing discharges and regenerative braking conditions.

Special Performance Test

The special performance test, as the name suggests, is defined only under certain conditions and/or environments that are encountered rather infrequently, and is also applicable under specific battery technologies.

Partial Discharge Test

This test provides a measure of the response of the battery pack to successive partial discharges, to identify any resulting capacity loss, and to verify the proper charging procedure for a partial DOD operation of the EV.

Standloss Test

This test measures the battery capacity loss when the battery is not used for an extended period of time. This test simulates the condition when the vehicle is not driven for a period of time, and the battery is not placed under charge. This standloss is primarily due to self-discharge, or other mechanisms that could result in permanent or semipermanent loss of battery pack capacity.

Sustained Hill-Climb Power Test

This test determines the DOD at which the battery will support a 6-minute discharge at approximately 45 W/kg before it is completely discharged. The test results are plotted with respect to time for which the 45 W/kg power can be sustained. This test provides a graphical determination of the maximum DOD at which the power can be provided for at least six minutes before the battery is fully discharged.

Thermal Performance Test

This test characterizes the effects of ambient temperature variation on the battery pack performance. The characteristics of the battery that are affected are in most cases, technology related. Thus the number and the types of charge and discharge cycles to be performed cannot be generalized for all battery types. The results of this test provide useful data to determine the need for battery thermal management or the allowable temperature range for a battery that may incorporate thermal management at a later stage.

Battery Vibration Test

This test characterizes the effect of long-term, road induced vibration and shock on the performance. This test characterizes the service life of candidate batteries. The intent is to either qualify the vibration durability of the battery or identify the design deficiencies that must be corrected.

Charge Completion on Optimization Test

This test establishes a charge algorithm or optimizes an existing algorithm for use in charging the battery pack in case the charge algorithm supplied by the battery manufacturer or developer does not adequately charge the battery, or a stable battery capacity is not obtained. Also, the charging method is a contributor to this problem.

Fast Charge Test

This test determines the fast charging capability of a battery by subjecting the battery to high charging rates, and determining the efficiency and other effects of accelerated charging. The USABC goal for fast charging is to return 40% of the state of charge (SOC) of the battery, starting from approximately 60% DOD in 15 minutes.

Performance Safety and Abuse Test

This test as the name suggests, characterizes the response of integrated battery systems to expected and worse-case accident and abuse situations. The information gained from these tests is used to qualify the vehicles as safe operating and accident/crash worthy.

ACCELERATED RELIABILITY TESTING OF ELECTRIC VEHICLES

Accelerated reliability tests of EVs are performed to provide several years of vehicle performance information in a short time—within a year. The tests are based on standard guidelines and the results of the tests provide a better understanding of EV performance.

These tests are conducted by operating the EVs in accelerated mileage modes that simulate typical fleet missions with the intent to obtain and analyze vehicle operation experiences within a relatively short period of one year. While these tests are conducted in an accelerated mode, they are always performed within the EV manufacturer's guidelines and do not void the manufacturer's warranty under any circumstances. In case the vehicle battery pack cannot be charged using a fast charge mode, the tests will be performed in accordance with no fast charging.

Each vehicle under the reliability test is tested for a period of one year. The vehicle is required to maintain the original components for the entire 25,000 miles testing period. The accelerated reliability tests may be carried out beyond the one-year period to gain additional vehicle knowledge.

The miles driven during the trip should be approximately 100 miles per day. Each driving cycle should be balanced with respect to the time required to charge so that the vehicle can be ready for the next test trip. The battery pack should not be discharged beyond 80% DOD unless recommended by the battery manufacturer. During the test, it is important to maintain the battery pack at 80% DOD.

The data gathered during the tests include:

- Average miles/charge by vehicle model (including potential range)
- Miles/charge by vehicle model (on a monthly or quarterly basis)
- Cumulative fleet mileage (on a monthly basis)

The charging pattern indicates when the vehicle is charged (time of the day), the duration of the charge (hours), and the magnitude of the charge (kWhr). In addition, the charging pattern also provides an assessment of the EV's impact on the utility system. If the time-of-use tariffs are available, the charging pattern also provides an accurate calculation of the cost of the energy consumed by the EV.

The charging data gathered during the accelerated tests includes:

- Number of charges per month
- Miles per charge
- Average charge time
- Energy consumed at on-peak rates
- Energy consumed at off-peak rates

This charge data obtained from the vehicle tests provides important vehicle information including:

- Miles/charge per month by vehicle and model
- Average recharge time by vehicle and model
- Average daily charging load profile by vehicle
- Average AC kWhr/mile by vehicle and by model

As part of the vehicle profiling data, it is useful to also determine the servicing man-hours by the vehicle model (both scheduled and unscheduled), vehicle availability, and downtime. The downtime of the vehicle may be further attributed to waiting for parts or downtime for maintenance.

The vehicle breakdown can also be attributed to on-road failure incidents. These will be breakdowns due to the vehicle failure while it is under test. On-road failure incidents also provide an estimate of repair costs per vehicle model down to the component level. Some the components that should be monitored are:

- Battery Pack
- Charging System
- Auxiliary System
- Traction Motor System
- Brakes and ABS System
- Tires
- Drivetrain and/or Transmission
- HVAC

In addition, the number of incidents/1,000 vehicle miles/per component is a good estimate of vehicle downtime. The inverse of the vehicle component downtime provides an estimate of the vehicle component reliability.

For the EV battery system, the following components should be monitored:

- Traction Battery
- Battery Modules
- High Voltage Battery Wiring
- Battery Pack Fan Filter
- Battery Fuse
- Battery Pack Disconnect Switch
- SOC Charge Gauge
- Battery Tray
- Battery Pack Fan
- Battery Pack Thermal System
- Battery Current Sensor
- Battery Temperature Sensor

For the EV charging system, the following components should be monitored:

- Onboard Battery Charger
- Onboard Charging Wiring
- Charger Unit Fan
- Charging Algorithm Interface Card
- Onboard Charging Port
- Charger Unit Fuse

For the vehicle traction motor system, the following components should be monitored:

- Traction Motor
- Traction Motor Filter
- Traction Motor Seal
- Traction Motor Throttle System
- Traction Motor Fan
- Traction Motor Cable
- Traction Motor Controller Connectors
- Traction Motor Mounting
- Traction Motor Hose
- Traction Motor Controller
- Traction Motor Wiring Harness
- Traction Motor System Cooling

For the vehicle drive train, the following components should be monitored:

- Axle
- CV Joint
- Differential/Rear Axle
- Rear Axle Seal
- Differential Mounting
- Parking Pawl
- Transmission/Transaxle
- Transmission/Transaxle Seal
- Transmission/Transaxle Mount
- Transmission/Transaxle Shifter

For the vehicle auxiliary system, the following components should be monitored:

- DC/DC Converter
- Auxiliary Battery
- Auxiliary Battery Fuse
- Backup Alarm
- Miscellaneous Belt System
- Heating System (Resistance/Fuel Fired)
- Miscellaneous Gages
- Power Steering Motor
- Power Steering Controller
- Power Steering Module
- Relay
- Warning Light
- HVAC
- Cooling System
- A/C Hose
- A/C Compressor
- Cooling System
- A/C Valve
- Heat Pump Reversing Valve
- A/C Motor Controller
- Ground Fault

For the vehicle brake system, the following components should be monitored:

- Brake Controller
- Power Brake Module
- Brake Rotor
- Brake Pads
- Regenerative Module
- Brake Drum
- Tire
- Wheel
- Tire Inflation Pressure Sensor
- ABS Sensor

Battery amortization costs should also be factored into the reliability test data. If the batteries in the battery pack are replaced during the period of the test, the battery replacement cost shall be divided for the miles traveled by the EV. This is to account for the vehicle cost per mile. This will provide a summary of the EV operating cost in $/mile or cents/mile based versus the energy consumed, maintenance, components, battery costs.

BATTERY CYCLE LIFE VERSUS PEAK POWER AND REST PERIOD

Two variables that affect the battery cycle life are the magnitude of the peak power demand during the battery driving profile and the during the battery rest periods (those times when the battery is neither under a charge nor under a discharge).

Differences in the peak power demand may be due to different driving characteristics of the driver or also from the difference in EV design. From the EV design standpoint, one vehicle may be a direct traction drive while the other may incorporate a transmission drive. Also selection of the battery pack rest period can be based on the convenience or on reduced battery charging station tariffs to take advantage of the off-peak charging.

In order to make a direct comparison, it is important to adjust the driving profiles such that the average power consumed remains the same. To accomplish this, the driving profile with the higher peak power incorporates a somewhat longer open-circuit or rest period within each driving profile. Evaluation of the influence of the timing of an eight-hour rest period is accomplished by inserting the rest period either immediately or after the completion of the driving profile discharge. Tests performed on the cycle life indicate that the battery arrives at its end-of-life as a result of degradation in the battery peak power handling capability.

Impact of Charge Method on Battery Cycle Life

The battery charge methods (a) constant current-constant voltage (CI/CV) with temperature compensated constant voltage and (b) three-step constant current (CI$_1$/CI$_2$/CI$_3$) with charge transitions at specific levels of the recharge factor, are tested before the battery reaches its end-of-life (end-of-discharge voltage).

The constant current (CI$_1$/CI$_2$/CI$_3$) charge method allows full recharging with a lower energy input and imposes a lower temperature stress on the battery module. It is expected that the full recharge will extend the battery cycle life.

Thermal Management of the Electric Vehicle Battery

The battery life and performance both are strongly affected by the operating temperature and the uniformity of the battery pack temperature. On the other hand, these temperatures can be modified by the battery thermal management system.

The thermal evaluation of the battery reveals that:

- Discharging the battery using a driving profile results in a significantly greater temperature rise than that found in a C/3 three-hour constant discharge.
- The battery thermal management system should account for the upper battery module to lower battery module pressure drop.
- The battery temperature rise, per unit energy discharge, is significantly greater than that of the flooded electrolyte battery. This is due to the lower heat capacity and greater heat generation. There is a poor heat dissipation of the starved-electrolyte battery.
- There is a large temperature difference between the electrodes and the battery cell due to the existence of air gaps between the electrode stack and the cell casing.
- Temperature increases are very sensitive to high-rate discharge, deep discharge, and excess overcharging.
- External cooling is less effective in the thermal management of the battery pack.

A major battery deficiency that inhibits a more rapid development of EVs is low specific energy. In the VRLA battery, this deficiency is at least partially attributable to the low utilization of the active material in batteries designed for deep discharging and long cycle life. The utilization of the positive active material is in order of 30%.

It is found that the factors that affect the battery utilization fall into three categories:

1. Application Conditions (e.g., discharge rate and operating temperature)
2. Battery design parameters (e.g., plate thickness, electrode porosity, electrolyte quantity, and comparison)
3. Intrinsic properties of the active material (e.g., active material composition, morphology, surface area, and crystallographic modification)

The use of thinner plates and increased porosity increases utilization but simultaneously reduces cycle life. Slight improvements in the utilization can be obtained by stirring the electrolyte and/or increasing the operating temperature.

Battery Test Recommendations

Based on the conclusions: (a) Peak power capability of the battery pack degrades more rapidly with cycle life than does the three-hour constant-current capacity. This peak power degradation determines the useful life under driving profile conditions. Consequently, the useful life with driving profiles that have a high peak power demand is much less than the life with profiles having a low peak power requirement. (b) Rest periods of up to eight hours between discharge and charge has little effect on battery life. (c) The three-step constant current ($CI_1/CI_2/CI_3$) charge method produces a lower temperature rise but does not yield increased cycle life over the constant-current, constant voltage (CI/CV) charge method recommended by the battery manufacturer. (d) Cooling of sealed starved-electrolyte lead-acid batteries is more difficult than for flooded-electrolyte batteries, particularly in those designs where the electrode battery assembly is not in contact with the battery casing walls. This is due to the greater heat generation and poorer thermal conduction in the sealed battery. (e) The design of the battery pack thermal management system can be significantly improved to improve the volumetric energy density of the system. (f) The most promising short-range approach to increasing the battery utilization of the active material in VRLA batteries is by the use of forced electrolyte circulation through the battery electrodes.

Modeling NiMH Batteries

A mathematical model for the NiMH cell was recently developed by Paxton and Newman. However, the model analysis was limited to the discharge behavior of the NiMH cell. While predicting an end-to-end performance of a battery pack in vehicle testing, it is important to account for the charge and overcharge reactions.

As part of the mathematical model development, it is important to incorporate the gassing phase of the NiMH battery. The NiMH cell can be broken down into the cathode (metal hydride) and anode (porous NiOOH), a separator and an electrolyte. The reaction at the anode is

$$NiOOH + H_2O + e^- \leftrightarrow Ni(OH)_2 + OH^-$$

With side reactions as

$$1/2O_2 + H_2O + 2e^- \leftrightarrow 2OH^-$$

At the cathode the reaction is

$$MH + OH^- \leftrightarrow M + H_2O + e^-$$

With the side reaction as

$$2OH^- \leftrightarrow 1/2O_2 + H_2O + 2e^-$$

When the equations for the anode and the cathode are combined they yield the overall NiMH reaction

$$NiOOH + MH \leftrightarrow Ni(OH)_2 + M$$

The side reactions at the anode and the cathode represent the oxygen recirculation reaction occuring inside the NiMH cell. During the charge phase, O_2 is generated at the Ni/electrolyte interface. The generated oxygen dissolves into the electrolyte. Once the electrolyte is saturated, the O_2 evolves in to the gas phase. This process of liquid to gas phase transition forms an internal oxygen cycle for the NiMH cell. The accumulation of oxygen in the gas phase results in pressure build up in the NiMH cell.

Some of the assumptions made include:

- The NiMH electrode consists of porous, cylinder with a substrate inside the electrode.
- The MH electrode consists of a uniform size, porous substrate.
- There is a continuous gas flow network with a uniform and a constant volume inside the NiMH cell.
- There is no contact between the active material and the oxygen gas phase as the electrolyte forms the bridge between the active material and the gas phase.
- Effects of convection of the electrolyte and the oxygen gas are negligible. Thus the movement of free O_2 occurs through the diffusion process.

Charge Acceptance of a cell is defined as the ratio of partial charge used by the electrochemical reaction to reverse the active materials to the total charge applied to the cell. The equation for charge acceptance may be expressed as

$$\text{Charge acceptance} = \Sigma a_{Ni} i_{n1} \, d \times dt \int I dt$$

where i_{n1} is the transfer current density, A/cm^2, I is the current density, A/cm^2. The charge acceptance is a function of the SOC and time.

Similarly SOC may be defined as

$$SOC = \Sigma a_{Ni} i_{n1} \, d \times dt / Q_{max}$$

where Q_{max} is the maximum theoretical charge capacity of the nickel electrode per unit projected area, C/cm^2. And DOD is expressed as

$$DOD = 1 + \Sigma a_{Ni} i_{n1} \, d \times dt / Qmax = 1 + SOC$$

Cell Pressure is expressed as

$$P = P_0 + (P^O{}_2 - P_0{}^O{}_2)$$

where P_0 is the initial (reference) cell pressure and is set to be zero. In the case of a sealed NiMH battery, the cell pressure results from the balance of oxygen generation, transport, and recombination. The cell pressure increases when the oxygen generation rate is higher than the oxygen recombination rate. This is a condition that occurs during battery charging and overcharging.

Polarization Resistance Model of NiMH Batteries

Polarization resistance associated with the NiMH battery is linear in nature at lower temperatures. The polarization resistance depends upon the SOC and is time dependent. This resistance is affected by electrode material composition, electrolyte level, electrode density, and electrode particle size. The polarization characteristic on discharge and the limiting current depends on the SOC of the electrode. The polarization resistance increases and the limiting current decreases as the SOC of the electrode decreases. Present at the hydride electrode, this resistance is ohmic in nature. The polarization resistance in the starved cell depends upon the amount of electrolyte in the cell. The resistance value decreases with increasing electrolyte quantity and the limiting current becomes lower. The polarization curve approaches a limiting current at higher current density.

Assuming an equivalent circuit for the interface consists of a double layer capacitor and a reaction resistance connected in parallel to the capacitance can also be calculated. The surface area of the electrode can be derived from the capacitance value. NiMH battery pack voltage (V_{pack}) maybe defined as

$$V_{pack} = OCV - IR - \Delta V$$

where the OCV is the open circuit voltage of the battery pack under no-load condition, I is the battery pack discharge current (A).

The polarization model for the NiMH may be defined as

$$\Delta RC \times d\Delta V/dt + \Delta V = I \, \Delta R$$
$$\text{since } I = I_0 + (d\Delta I/dt) \times t$$
$$\text{Therefore, } \Delta V = \Delta R \, [(I_0 - \Delta RC \times (d\Delta I/dt)(1 - exp(-t/\Delta RC)) + (d\Delta I/dt) \times t]$$

For example, assume that the battery pack is at 7% DOD and the OCV for the NiMH cell is 1.359 V/cell, R_{batt} is 0.225 Ω, ΔR is 0.059 Ω. Therefore, using the above equation, ΔRC = 21.3 seconds.

The polarization resistance increases with decreasing temperature. The temperature dependence indicates that it may be associated with the resistance of the KOH electrolyte. Thus making the polarization resistance time dependent and SOC dependent.

The Thermal and Electrochemical Coupled Model

The variation of temperature and the distribution of temperature inside the battery may be simulated using a thermo-electrochemical model.

Temperature is one factor that affects the battery performance, life, and reliability of operation. Battery physiochemical properties are greatly strong functions of temperature. The equilibrium pressure of hydrogen absorption-desorption, affects the open-circuit potential of the metal hydride electrode and hence the NiMH. Battery capacity losses occur at low temperatures due to high internal resistance and at high temperatures due to rapid self-discharge. Thus making the operating temperature range essential for a battery pack to achieve optimal performance. Battery life can be improved by balanced utilization of active materials, which requires a highly uniform temperature profile inside the battery to avoid the localized degradation. The battery temperature may increase due to self-accelerating characteristics of exothermic reactions resulting in the evolution of oxygen eventually leading to a thermal runaway. In order for the battery pack to operate optimally, it is essential to maintain an optimal operating range and a high uniformity in the internal temperature distribution of the battery.

The battery thermal model is based on the thermal energy balance over the representative volume in the battery. The differential equation that describes the temperature distribution in the battery is expressed as

$$\rho C_p(\partial T/\partial t + v \times \nabla T) = \nabla \times \lambda \nabla T + q$$

where ρ is the volume-averaged density, C_p is the specific heat, λ is the heat conductivity of the battery volume under study, v is the velocity vector, q is the volumetric heat generation rate. Thus the temperature associated with convection energy may be represented as $\rho C_p v \times \nabla T$, temperature associated with the energy accumulation may be represented as $\rho C_p \nabla T/\partial t$. The temperature associated with the conduction energy is represented as $\nabla \times \lambda \partial T$.

The thermal model of the traction battery can be thermally and electrochemically coupled or decoupled, depending on how the heat generation is analyzed. During the battery operation, the heat generation rate depends on the cell temperature and also on the charge and discharge profiles. A coupled model of the traction battery uses newly produced current and potential data to determine the heat generation rate and the temperature distribution. The heat generation rate and the temperature distribution of the battery may be used to determine the battery current and the battery potential. The decoupled model on the other hand uses empirical relations at constant temperatures. This decoupled model is accurate only when the battery performance is insensitive to temperature variations.

The heat generation rate depends on the thermodynamic properties of the reactions in a cell, the potential-current characteristics of the cell and the rates of charge and discharge. Under constant pressure conditions, Bernardi suggests that the rate of heat generation is given by

$$Q = \Sigma I_j(U_j - T\partial U_j/\partial T) - IV + \text{enthalpy of mixing} + \text{change of phase}$$

where I_j is the volumetric partial reaction resulting from the electrode reaction j, U_j is the corresponding open-circuit voltage (OCV), I is the total current per unit volume (A/cm^3), and V is the cell potential. The first term on the right hand side in the equation for Q represents the enthalpy of the charge-transfer reactions. The second term represents the electrical work performed by the battery. The third term on the right hand side in the equation represents the enthalpy of mixing or the heat effect associated with concentration gradients developed in the traction battery cell. The equation for Q assumes that the temperature of the cell is held constant. This is a good approximation when the traction battery cell is thin and there is a large heat exchange due to convection and conduction. However, in case of a battery, multiple electrode reactions occur simultaneously. The equation for Q can be used to determine the partial current of each electrode reaction.

In case of NiMH and Li-ion batteries, the OCV is a function of the SOC, which in turn depends on the rate of diffusion. During high rate electrochemical reactions, the general energy balance, as expressed by Rao and Newman represent the rate of heat generation as

$$Q = -1/V_c \int \Sigma\, a_j i_j(U_j - T\partial U_j/\partial T)\ dV - IV$$

where a_j is the specific active surface area for electrode reaction, i_j is the transfer current density, and U_j is the OCV of the reaction. This equation represents the heat generated by the traction battery when the cells undergo relaxation under dynamic conditions.

The thermal runaway condition is triggered by the hottest spot in the cell. Thus a constant temperature model is not a successful prediction of the heat generated during an exothermic reaction in a traction battery.

In case of the lumped thermal model, the heat generated may be approximated to

$$\rho C_p(\partial T/\partial t) = \nabla \times \lambda \nabla T + q$$

where the heat generated due to convection is neglected. The thermal conductivity λ is dependent of the structure of the battery cell.

For single cells with negligible thickness

$$Bi = hL/\lambda \ll 1$$

where h is the convective heat-transfer coefficient, and L is the thickness, and λ is the thermal conductivity of the cell.

Diffusion Coefficient and Ionic Conductivity

Using the Arrhenius equation the diffusion coefficient is expressed as

$$\Phi = \Phi_{ref} \exp [E_{act}/R \times (1/T_{ref} - 1/T)]$$

where Φ is the diffusion coefficient, conductivity of the electrolyte, exchange current density of an electrode reaction, E_{act} is the activation energy.

In addition the OCV, V_{ocv} is approximated as a linear function of temperature

$$V_{ocv} = V_{ref} + (T - T_{ref})\partial V_{ocv}/\partial T$$

where the V_{ref} is the reference voltage, T_{ref} is the reference temperature.

$$Q = hA_c(T - T_a)/V_c$$

where V_c is the cell volume, A_c is the surface area through which heat is removed from the traction battery, and T_a is the ambient temperature of the surrounding atmosphere.

Thermal Effects During CC Charging

As the NiMH cell is charged, the heat-transfer coefficient decreases, the cell potential increases gradually, and the potential peak becomes more prominent. The cell temperature increases with time, which indicates that the cell charging is an exothermic reaction. A larger heat-transfer coefficient corresponds to a larger rate of heat dissipation resulting in a smaller temperature rise. When the heat transfer coefficient is larger than $5\,W/m^2K$, the cell temperature rise is less than 5°C when the charge input is less than 90% of the nominal cell capacity. After this stage, the cell temperature increases to almost 64°C at 120% charge input when the heat-transfer coefficient equals $5\,W/m^2K$. As the heat-transfer coefficient increases, the final cell temperature decreases. A less than 10°C increase from the original temperature is observed when the heat-transfer coefficient is as large as $25\,W/m^2K$. The cell temperature exceeds the safety limit for the aqueous cell (80°C) when the cell is charged under adiabatic conditions. Thus showing the need for thermal management of the NiMH batteries.

The total heat generation rate increases slowly before 90% charge input, thereafter the heat generation rate increases rapidly and finally reaches a steady state and matches the trend in the temperature variation closely. The thermal environmental condition has little effect on the heat generation rate. When the battery is charged at a constant current, the overall reaction current is fixed. Slight variations in the total heat generation rate result from the difference in the ratio of the primary to secondary reaction rates.

The MH formation and Joule's Law heating effect contribute to the total primary heat generation rate. This is cause-effect relationship as during the initial reaction, the heat effect is due to MH formation, while the oxygen reactions are insig. ificant and the Joule's Law heating effect is negligible. As time progresses, heat absorbed by the primary reactions decreases because of the reduction in their enthalpy potentials, and the heat generated from the metal hydride formation remains constant. This effect in turn increases the total heat generation rate.

When the traction battery cell is being charged and the oxygen reaction becomes significant, the total heat generation rate increases dramatically since the enthalpy potentials of oxygen reactions are large and the heat absorbed by the primary reactions is negligibly small. Under overcharge condition at the rate of 1C, all the current applied to the cell is used to generate oxygen at the anode and only a small portion of oxygen can be reduced at the cathode. Thus the primary reaction at the MH electrode accounts for a large portion of the applied current by the heat effect due to MH formation. The net heat generation rate is large, but the steady state is reached since the reaction rate is constant.

Initially the oxygen reaction is negligible, and the current applied to the cell is used to convert the active materials from the discharged to the charged state. Once the charge input exceeds 90% of the nominal cell capacity, the oxygen reactions become significant. While the oxygen reaction current increases, the primary reaction current decreases when the total current remains the same. Charge acceptance, defined as the ratio of primary reaction current to the total current, exactly follows the primary reaction current with a maximum value of 100%. Thus the heat dissipation rate affects the charge acceptance once the charge input applied exceeds 100% of the nominal capacity. The worse the heat dissipation, the smaller is the charge acceptance by the traction battery.

During the CC charge mode, the float charge at constant voltage is frequently applied to the NiMH battery. The current produced due to oxygen reactions is negligible when the charge input is less than 60% of nominal cell capacity. In case the battery is not sufficiently cooled, early release of oxygen occurs resulting in higher cell temperatures.

The current, due to primary reactions, decreases very quickly during the first 10% of the nominal cell capacity because the surface overpotential drops rapidly due to increase in the electrode's OCV during charging. The primary reaction current is also affected by the rate of battery cooling. The higher the heat dissipation rate, the lower is the cell temperature, and thus the lower is the reaction current. The reaction current continues to increase even after the initial drop in value, indicating that the reactions are superficial in nature. When both the primary and the oxygen reaction currents are small, the total reaction current is small.

As mentioned, cell temperature is strongly affected by the cooling condition. As the cell temperature increases due to poor cooling conditions, it decreases continuously after an initial rise in temperature. This drop in temperature depends on the heat transfer coefficient. The larger the heat transfer the coefficient more rapid is the temperature drop. The decrease in the traction battery temperature tapers off as the battery approaches its fully charged state, and the oxygen reactions become dominant. The cell temperature approaches a constant value at high cooling rates, the cell virtually undergoes thermal runaway at low cooling rates.

In addition to temperature rise, the oxygen reactions contribute to a cell pressure build-up. The sooner the oxygen reactions take place, the more the cell pressure builds up. While the cell temperature is low and the heat dissipation rate is high, the traction battery cell can be maintained at a low level.

Electric Vehicle Performance Model

The EV system performance is based on the integral performance of the individual components. The battery is electrically connected to the inverter/controller on board the EV. Auxiliary loads and auxiliary power units are directly connected to the traction battery. The motor and inverter/converter are also connected via direct electrical connection. A traction motor is mechanically connected to the traction gear transmission and to the wheels. Losses amongst the individual components (e.g., transmission losses) are not modeled. Friction and aerodynamic losses in rotating components are associated with the particular component but are collectively reflected in efficiency maps.

Driveshaft Power

Driveshaft power is defined as the total power at a given time required at the driveshaft of the vehicle to move the vehicle from the speed at one time step to the speed at the next time step (i.e., v_{t-1} to v_t). The driveshaft power is calculated using

$$P_d(t) = P_{acc} + P_{grade} + P_{aero} + P_{rolling} + P_{bearing}$$

$$P_{acc} = (M_c \times dv(t)/dt) \times v(t)$$

where P_{acc} is the inertia power associated with acceleration and deceleration, P_{grade} is the power associated with gravitational acceleration, P_{aero} is the power associated with aerodynamic drag, $P_{rolling}$ is the power associated with rolling resistance, and $P_{bearing}$ is the power associated with wheel bearing loss.

Each of the components associated with the total power may be expressed as

$$P_{acc} = (M_e \times dv(t)/dt) \times v(t)$$

$$P_{grade} = W \times \sin(\theta) \times v(t)$$

$$P_{aero} = [1/2\ \rho C_d(\gamma) \times Av_r(t)^2] \times v(t)$$

$$P_{rolling} = [(C_0 + C_1 v(t) + C_2 v(t)^2 + C_3 v(t)^3) \times W] \times v(t)$$

$$P_{bearing} = \tau_B \times s_W(t)$$

where M_e is the mass equivalent and includes the rotational inertia of 3% of the weight of the vehicle, g is the acceleration due to gravity, ρ is the density of air, $C_d(\gamma)$ is the aerodynamic drag coefficient as a function of wind with a yaw angle of γ, A is the frontal area of the vehicle, $v(t)$ is the vehicle speed as a function of time, $v_r(t)$ is the relative wind speed in the direction of travel as a function of time, W is the weight of the vehicle, τ_B is the wheel bearing torque drag, $s_W(t)$ is the wheel speed as a function of time, and θ is the road grade angle measured with respect to the horizontal. C_0, C_1, C_2, and C_3 are the coefficients of rolling resistance.

In case the power at the driveshaft of the vehicle is negative for the time interval from v_{t-1} to v_t, the power at the driveshaft is calculated as

$$P_d(t) = P_{acc} + P_{grade}$$

In this case the power is absorbed to slow the vehicle. The driveshaft torque τ_d is calculated from the vehicle speed $v(t)$ and the tire rolling radius r.

Thus

$$S_W = v(t)/2\pi r \text{ and } \tau_d = P_d/S_W$$

In addition, τ_d may be expressed as $f_r \cdot (P_d/S_W)$ where f_r is the fraction of torque available for regenerative braking. And $(1 - f_r)$ is the fraction of the torque absorbed by the vehicle's friction brakes.

Vehicle Tire Limit is the maximum possible wheel force for acceleration, F_W acceleration and deceleration, F_W deceleration is calculated as

$$F_W \text{ acceleration} = \mu_0 \times W \times f_d \text{ and } F_W \text{ deceleration} = \mu_0 \times W$$

where μ_0 is the coefficient of static friction between the road and the EV tires, W is the weight of the EV, and f_d is the fraction of the weight supported by the wheels of the EV.

Electric Vehicle Motor

The power of the motor P_m is calculated as

$$P_m = P_d \times \eta_{tr} \text{ for } P_d > 0 \text{ (regenerative braking quadrant of operation)}$$

and

$$P_m = P_d/\eta t \text{ for } Pd > 0 \text{ (driving quadrant of operation)}$$

The motor speed S_m and the motor torque τ_m is then calculated from the transmission gear ratio r_{gear}, the transmission driveshaft speed and the previously calculated motor power P_m.

The motor speed is expressed as $S_m = s_d \times r_{gear}$, and the motor torque is expressed as $\tau_m = P_m/S_m$

Maximum Motor Torque

The maximum motor torque check is performed by comparing the motor torque τ_m to the maximum motor torque $\tau_{m\ max}$.

In case the maximum motor torque envelope is exceeded, τ_m is set equal to $\tau_{m\ max}$. The solution for the relationship is found by iteration of $v(t)$.

$$P_{m\ max} = P_d/\eta_{tr}$$

where $P_{m\ max}$ is a function of motor speed, P_d is a function of the EV speed, and η_{tr} is a function of the driveshaft torque and speed.

Under acceleration, the solution provides the fastest speed possible within the maximum motor torque envelope. Under deceleration, the maximum amount of regenerated power is returned back to the battery.

Inverter/System Controller

Based on the motor power P_m and the motor efficiency at the current operating point η_m the power at the motor/system controller interface P_i, is calculated using the equation

$$P_i = P_m \times \eta_m \text{ for } P_m < 0 \text{ (based on the regenerative braking}$$
$$\text{quadrant of operation)}$$

and

$$P_i = P_m/\eta_m \text{ for } P_m > 0$$

Traction Battery

The traction battery characteristics are calculated using the inverter/system controller power P_i and the inverter/system controller efficiency η_i. The traction power at the battery/inverter system controller P_{batt} is calculated using the following relationship

$$P_{batt} = P_i \times \eta_i \text{ for } Pi < 0 \text{ (for the regenerative braking operation)}$$

$$P_{batt} = P_i/\eta_i \text{ for } P_i > 0 \text{ (for the driving operation)}$$

The total battery power P_{batt} is determined by adding the power contributed by the auxiliary power unit P_{APU}, and the auxiliary loads P_{aux} is based on the following equation

$$P_{batt} = P_{APU} + P_{aux} + P_{tract}$$

Battery power and current leaving the battery is considered to be positive (discharge condition) and current entering the battery is considered to be negative (charge condition). Using this battery convention, the P_{APU} is generally negative and P_{aux} is generally positive.

The total battery power P_{batt} is determined from the battery characteristics (OCV, resistance dV/dI_b at a given DOD)

$$P_{batt} = dV/dI \times I_b^2 + OCV \times I_b$$

which is given by the following expression

$$[-OCV + (OCV^2 - 4 \times dV/dI \times P_{batt})^{1/2}]2 \times dV/dI$$

The battery voltage under load, V_{batt} is expressed as

$$V_{batt} = P_{batt}/I_b$$

Since it can be assumed that most components in the EV are connected in parallel, the current values for the auxiliary power unit, auxiliary loads, and powertrain may be determined as

$$I_{APU} = P_{APU}/V_{batt}$$

$$I_{aux} = P_{batt}/V_{batt} \text{ and}$$

$$I_{tract} = P_{batt}/V_{batt}$$

Minimum Battery Voltage

The battery pack voltage is compared to the minimum voltage and is expressed as V_{min}. If $V_{batt} < V_{min}$ then V_{batt} is set to the value of V_{min}, and the maximum battery current at this voltage is found from the relationship

$$I_{Vmin} = (V_{min} - OCV)/dV/dI$$

The remaining current and power at the minimum battery voltage available to the powertrain is calculated as

$$I_{Vmin, tract} = I_{Vmin} - (P_{APU}/V_B) - (P_{aux}/V_B)$$

$$P_{tract,Vmin} = I_{Vmin,tract} \times V_{min}$$

Alternately, $P_{tract,Vmin}$ may be expressed as $Pd/\eta_{trans} \cdot \eta_m \cdot \eta_i$ where Pd is a function of the vehicle speed and η_{trans}, η_m and η_m are the functions of the driveshaft torque τ_d and vehicle speed v(t).

The maximum vehicle speed under the minimum battery voltage conditions is expressed with maximum current I_{max}. I_{max} is a function of motor speed S_m. If the maximum current for the calculated motor speed $I_{tract} > I_{max}$ then I_{tract} is set equal to $I_{max,tract}$.

The V_{batt} is determined by

$$V_{batt} = OCV - (I_{max} + I_{APU} + I_{aux}) \times dV/dI$$

The battery voltage, V_{batt} is compared to the minimum voltage, V_{min}. If V_{batt} is less than V_{min}, then V_{batt} set equal to V_{min}. If V_{batt} is greater than V_{min}, the maximum power available to the powertrain P_{tract} and the currents required for auxiliary loads, auxiliary power unit, and total battery current I_{batt} is calculated based on the following relationships.

$$P_{tract} = V_{min} \times I_{max}$$

$$I_{aux} = P_{aux}/V_{min}$$

$$I_{APU} = P_{APU}/V_{min}$$

$$I_B = I_{max} + I_{aux} + I_{APU}$$

$$P_{tract}, V_{min} = P_d/(\eta \times \eta_m \times \eta_l)$$

where P_d is a function of vehicle speed and η, η_m, and η_i are functions of the driveshaft torque τ_d and vehicle speed $v(t)$.

Auxiliary Power Unit

In a series hybrid configuration and the battery DOD at which the EV switches into and out of the hybrid mode of operation and variable power generator output power, are modeled as follows.

The generator power P_{gen} is based on the minimum and maximum values selected by the user and the average power required to propel the vehicle on the driving cycle during the previous time period. The average power P_{av} is calculated by averaging the traction power (power input to the inverter/system controller) over a specified time period Δt_{gen} is expressed as

$$P_{average} = \Sigma p_{tract}/\Delta t_{gen}$$

P_{gen} is then calculated by $P_{gen} = f_g \times P_{average}$ where f_g is the user input factor and P_{gen} is subject to the maximum and minimum specified power allowed. The generated power is expressed as

$$P_{gen} = P_{gen(min)} \text{ if } P_{gen} < P_{gen(min)} \text{ and } P_{gen} = P_{gen(max)} \text{ if } P_{gen} > P_{gen(max)}$$

The efficiency of the generator η_{gen} is expressed as the function of ratio of P with respect to P_{max}.

The output power of the engine is expressed as

$$P_{eng} = P_{gen}/\eta_{gen}$$

where P_{gen} is the power out of the engine and η_{gen} is the generator efficiency.

The fuel used (Δ_{fu}) at each increment Δt is calculated based on the expressions

$$\Delta_{fu} = brake_{fc} \times P_{eng} \times \Delta t_i/3600$$

$$\Delta_{fu} = brake_{fc} \times P_{eng} \times \Delta t_i/3600 \times /(\rho_f)$$

where $brake_{fc}$ is a brake specific fuel consumption (gm/kWhr) of the engine, and ρ_f is the density of the fuel.

The emissions of

$$HC, CO, \text{ and } NO_x = (gm/kWhr)\ HC, CO, NO_x \times P_{eng} \times t_i$$

where (gm/kWhr) HC, CO, and NO_x are the specific emissions of the engine.

The emissions downstream for the catalytic converter are given by

$$e(HC, CO, NO_x)catalytic =$$
$$e(HC, CO, NO_x)_{eng} \times [1 - \eta_F(HC, CO, NO_x)catalytic$$

where $\eta_F(HC, CO, NO_x)$ are the conversion efficiencies of the catalyst for the pollutants HC, CO, and NO_x. The efficiency of the catalyst depends on its temperature. It increases with time when the engine is turned on and decreases with time after the engine is turned off. These effects are modeled by expressing the catalyst efficiency as

$$\eta_c/\eta_0(HC, CO, NO_x) = 1 - e^{-(te/tc)}$$

t_e is the effective time the engine is on, t_c is the characteristic time for warm-up of the catalyst, and $\eta_0(HC, CO, NO_x)$ are the steady state conversion efficiencies of the catalyst for HC, CO, and NO_x. The effective engine on-time is calculated as follows

$$\text{engine on-time } t_e = t_e + [M1]\Delta t_{on}$$

$$\text{engine off-time } t_c = t_c - a \times \Delta t_{off}$$

where a is a factor accounting for the difference between the heat-up and cool-down times of the catalyst. The maximum value of t_c permitted is that corresponding to η_c equal to 99% of η_0 for each of the pollutants.

Battery Capacity and %DOD

The Ahr in and out of the battery can be summed by the average Ahr through the battery. The quantities are then divided by time intervals to calculate the average charge, discharge, and net current.

$$I_{avg,charge} = \Sigma(I_{charge} \times \Delta t_{charge})/\Sigma \Delta t_{charge}$$

$$I_{avg,discharge} = \Sigma(I_{discharge} \times \Delta t_{discharge})/\Sigma \Delta t_{discharge}$$

$$\text{where } Ahr_{net}\Delta t = \Sigma(I_{batt} \times \Delta t)$$

The average battery current $I_{average}$ is used to calculate the available battery capacity from the Peukert constants using the constants from the battery file. The equation is expressed as

$$Ahr_{capacity} = a \times I_{average}^b$$

where a and b are constants.

The %DOD at each time increment is expressed as the equation

$$\%DOD = Ahr_{actual}/Ahr_{capacity}$$

Battery Scaling

Battery scaling methodology is straightforward. After the user enters a modified battery Ahr capacity, the Peukert relationship. The dV/dI with respect to %DOD, battery weight is corrected based on the following relationships:

$Ahr_{capacity}(new) = Ahr_{capacity}$, a = new Peukert constant, (dV/dI) is the resistance, and W_{module} is the battery module weight.

$$DV/dI_{batt} = dV/dI_{batt} \times Ahr_{capacity(new)}/Ahr_{capacity}$$

$$a = b \times (Ahr_{capacity}(new)/Ahr_{capacity})^{1-b}$$

$$W_{module} = W_{module} \times Ahr_{capacity}(new)/Ahr_{capacity}$$

Energy

Based on the power of each component over their appropriate time intervals, the energy may be expressed as

$$E_{batt} = \Sigma(P_{batt} \times \Delta t) \text{ when } P_{batt} > 0$$

$$E_{APU} = \Sigma(P_{APU} \times \Delta t) \text{ when } P_{APU} < 0 \text{ and } P_{APU} > 0$$

$$E_{aux} = \Sigma(P_{aux} \times \Delta t) \text{ when } P_{aux} > 0$$

$$E_{tract, driving} = \Sigma(P_{batt, tract} \times \Delta t) \text{ when } P_{batt, tract} > 0$$

$$E_{tract, regen} = \Sigma(P_{batt, tract} \times \Delta t) \text{ when } P_{batt, tract} < 0$$

SAFETY REQUIREMENTS FOR ELECTRIC VEHICLE BATTERIES

During 1993, the National Highway Transportation and Safety Association (NHTSA) conducted research and testing of two converted EVs.

The EVs were subjected to crash tests with Pb-acid batteries in two different configurations. The crash test was conducted at approximately 45 to 50 mph into a fixed barrier. The results of the tests demonstrated significant damage to the front of the battery pack with large quantity of electrolyte spillage. In one of the two vehicles the electrolyte spilled exceeded 12 ltrs. In addition, several high voltage arcs were also observed under the hood of the EV during the crash.

Based on the results of the crash tests and the increasing demand for EVs, the NHTSA has reexamined the safety concerns of EVs especially with respect to battery electrolyte spillage and electric shock hazard.

When a vehicle is converted into an alternate-fuel vehicle, before the first sale to the ultimate consumer, the converter is considered to be identical to the vehicle manufacturer. The converter must certify that the vehicle still complies with all applicable standards. This includes fuel system integrity standards applicable to the alternate fuel. For example, if a converter before the first sale converted the IC engine vehicle to an EV, then the converter will need to certify to the spillage requirements or recall and remedy the noncompliant vehicles by replacement or repair.

In the case of a crash, the performance requirements address electrolyte spillage in a crash or rollover—thus limiting the amount of allowable spillage through a performance test. After a barrier crash test there would be no more than (a) one ounce (28 gms) by weight of liquid fuel loss from the time of the barrier impact until the vehicle motion has ceased; (b) five ounces (142 gms) during the next five minutes; and (c) one ounce (28 gms) per minute during the next 25 minutes. These requirements apply to all vehicles of 100,000 lbs. (4,536 kg) weight or less when subjected to a 30 mph (48 kmph) lateral or 30 mph (48 kmph) rear moving barrier crash test. For school buses with a weight greater than 10,000 lbs. (4,536 kg), requires a 30 mph (48 kmph) moving barrier impact at any point from any angle on the bus with the same allowable fuel loss.

Some of the questions that arise include:

- What is the appropriate amount of electrolyte spillage to allow after a crash or rollover test?
- Is this a factor based on the number of or type of batteries onboard the EV?
- Does it depend on the number or type of batteries onboard the EV or whether the spillage occurs inside or outside the passenger compartment or cargo areas of the vehicle?
- Is there an amount that can be considered to be a no-spillage condition?

- Is there a threshold temperature above which spillage requirements are needed?
- Are there additional performance requirements that should be considered in addressing the safety hazards of the EV?
- Is there an optimal battery pack configuration based on the vehicle weight?
- Is there a location of the battery pack that will minimize their damage in a crash or rollover?
- What are the requirements that should be placed on battery pack design, construction, or testing? Should battery packs be dual-walled compartments?
- In the case of hybrid electric vehicles (HEVs), are there additional safety problems presented relative to electrolyte spillage or electric shock?
- Are there additional safety problems associated with the HEVs when both fuel sources are being utilized?
- The IC engine drivetrain has a forward, neutral, and reverse gear transmission. In the case of some EVs, a similar transmission is not used. Is there a safety problem associated with inadvertent starting and unwanted vehicle motion?

Appendix A FUEL CELL PROCESSING TECHNOLOGY FOR TRANSPORTATION APPLICATIONS: STATUS AND PROSPECTS

There is little dispute that automotive designs based on fuel cell technology have the potential to provide high efficiency while dramatically reducing tailpipe emissions. A fuel cell is a propulsion system that can replace the internal combustion (IC) engine. Fuel cells have several features that make them suitable for automotive applications including:

- **Clean Air** Fuel cell vehicles running on hydrogen fuel produce no pollution or greenhouse gases. They are zero-emission vehicles (ZEVs) whose only output is water vapor.
- **High Fuel Efficiency** Fuel cells convert chemical energy directly into electricity, so they tend to lose less energy, to waste heat, and have efficiencies two to three tines higher than the IC engines.
- **Quiet Operation** Although the vehicle auxiliary systems produce noise, the fuel cells themselves do not. Overall the fuel cell vehicle is quieter than the conventional vehicle.
- **Good Vehicle Performance** Fuel cell vehicle performance is equal to that of the conventional IC engine. In contrast to the battery-powered electric vehicle (EV), with limited range and relatively long recharging times, the fuel cell vehicles are expected to provide long distance mileage, over 250 miles between refueling, and should be able to refuel in a matter of minutes.
- **Long Operating Life and Low Maintenance** The high temperature operation and friction caused by the moving parts in an IC engine cause significant wear and require high maintenance. Fuel

cells, on the other hand operate under low temperature conditions and do not require additional moving parts.

Thus fuel cell vehicles combine the quiet operation and environmental advantages of the battery-powered electric vehicles with the long range and rapid refueling time of the conventional IC engine. Another advantage of the fuel cell vehicle is that it can be powered by hydrogen rich mixture. Although fuel cells run on H_2, other fuels can be stored on board the vehicle and reformed into a hydrogen rich gas. This reforming process combines steam with the fuels such as gasoline, methanol, or natural gas to produce H_2, CO_2, and small amounts of other gases. As gasoline reformers continue to be inefficient and natural gas reformers are too bulky, methanol based cells are suitable alternatives to hydrogen.

Significant developments in the past four years have enhanced the prospects for fuel cell applications in transportation. Leading from these developments, the following conclusions can be arrived at:

- The successful direction of the U.S. transportation fuel cell program has resulted in rapid increase in activity in Japan and Europe with participation of major industry interests in automotive developments.
- Proton Exchange Technology (PET) based fuel stack technology has progressed rapidly over the last 4 years to achieve a 60% reduction in size. Improved engineering design based on already demonstrated technology will bring stack weight and volume within the envelope established by the U.S. automakers for advanced vehicles.
- Over the last 4 years, the underlying factors leading to the extraordinary high cost (greater than 3,000 $/kW) of the fuel cell stacks have been identified and gradually reduced. Significant progress has been made in the areas of reduction of required platinum (Pt) to levels no higher than used in today's catalytic converters, further reduction in membrane costs and overall reduction in materials consumption. Existing technology has the capability to achieve costs of approximately $250/kW solely by the achievement of even modest automotive production volumes. Further implementation of technology improvements have demonstrated that the fuel cell stacks can cost less than $50 to $75/kW.
- Significant progress has been made in fuel processor development, though the technology is less mature than fuel cell stacks. The feasibility of low-cost, multifuel reformers has been successfully demonstrated. Thus the initial introduction of automotive fuel

cells based on gasoline can be facilitated by use of gasoline as the fuel.

- Operation of the fuel cell stack based on ethanol or alternative renewables provides an alternate fueling option. Fuel processor subsystems now cost $50/kW for automotive applications.

Despite the significant developments in the fuel cell technology, capturing the results for commercial purposes will require significant investments at the system level. Testing and manufacturing systems will need to scale up significantly.

The next generation advanced vehicle program has a set of objectives that has called for approaching 80 MPG, with a full-sized vehicle without compromises in emissions, range, driveability, cost, and safety. These are a very challenging set of requirements. In addition, there are multiple technology paths that should be considered for the development of advanced vehicles that meet the vehicle design objective. These include hybrid vehicles with compression ignition and gas turbine engines.

Fuel cell technology is developing rapidly to meet all of the advanced vehicle requirements including:

- Very high traction efficiency
- Negligible or low tailpipe emissions
- Multifuel operation capability with use of a reformer
- Low noise/vibration during driving
- Potential for meeting required automotive packaging/cost requirements

In particular, the combination of very high efficiency and negligible emissions provides the fuel cell vehicle with an advantage to simultaneously address major energy/environmental driver concerns.

The Department of Energy's (DOE) fuel cell bus has demonstrated extremely low emissions. Measurements have also verified zero particulates and only trace NO_x and hydrocarbons. These measurements need to be verified over the life of the vehicle, but it can be expected that the negligible emissions associated with fuel cells will essentially reduce related automotive emissions. Under the most stringent standard, the fuel cell vehicle will qualify as a zero emission vehicle.

The high efficiency of fuel cells will result in a major increase, at least by a factor of 2, in the fuel economy as compared to the conventional IC engine technology. At the rated power, fuel cell system level efficiencies are as high as (or higher than) any other identified automotive design options, as illustrated in Table A–1. Under part-load conditions,

Table A–1   Fuel cell technology efficiency ratings.

System Type	Peak System Efficiency (%)
PEM on hydrogen	65
PEM with reformer-methanol	56
PEM with reformer-gasoline	45
Gasoline IC engine	32
Diesel IC engine	45

which often determine driving cycle fuel economy, fuel cell technology has an additional advantage. The high peak and part load efficiency characteristics will allow the fuel cell vehicles to achieve demanding advanced vehicle design goals without radical design compromises.

Since the fuel cell system design program was initiated, there have been rapid developments in the technology including:

- Reductions in the catalyst quantity by over an order of magnitude to levels not significantly different from those employed in the automotive catalytic converters
- Significant testing of fuel processing technologies consistent with operation of multiple fuels, including ethanol, methanol, methane, and gasoline
- Dramatic reductions in size and weight of fuel cell stacks, with reduction in the design parameters by four-fold
- Projections (based on hardware testing) of high efficiency, over 50% and extremely low emissions on both hydrogen and reformed fuel
- Significant improvements in basic electrochemical performance have resulted in a 50% increase of fuel cell power density and efficiency

Companies including Ansaldo, Siemens, Mitsubishi, Fuji Electric, Toshiba, Mitsubishi Electric, Sanyo Electric, Toyota, and Honda have made significant development announcements in both Europe and Asia.

In Europe, under a new framework program, it is estimated that proton exchange membrane (PEM) fuel cells research for transportation are receiving about 60 million ($15 million per year). The program participants are both automotive manufacturers and suppliers. The manufacturers include Daimler-Chrysler, BMW, Volkswagen, Volvo, Renault, and Peugeot. While the suppliers include Johnson Matthey, Siemens, Ansaldo, Rolls-Royce, CJB, and DeNora.

In Japan, the polymer electrolyte fuel cells have been in development as part of the Phase I for the past six years. In the new Phase II, the fuel cell stacks are targeting the power stacks of several tens of kW. There are now at least seven companies in Japan that have operated PEM fuel cell stacks with power ratings ranging from 1 kW to 5 kW. The reduction of the platinum (Pt) has been successful and small stacks have been operated successfully for over 8,000 hours exceeding the advanced vehicle life design goals of 5,000 hours.

System level analyses indicate that one of the critical requirements for automotive applications is a fuel cell stack with power density of 1.2 kg/kW on a weight basis and 1.5 ltr/kW on a volume basis. There are several new design aspects of the fuel cell design that will allow for the next generation developments to take place. These include the weight and volume aspects of the fuel cell stack. The fuel cell stack is composed of electrodes (called the membrane-electrode-assembly [MEA]), cooling plates to remove the heat being produced, flow fields to distribute reactant gases across the electrode surfaces. In addition, there are structural members to clamp and manifold the stack. In the 5 kW polymer electrolyte fuel cell stack, the active cell weighs more than half of the weight of the fuel cell. The next heavy component is the endplate followed by the humidification cell and the cooling cell. In the past designs of the fuel cells, the active cell represents just over a fourth of the total fuel cell weight. Similar observations apply to the volume. Publicly demonstrated stacks since the past few years have already achieved greater than a 60% reduction in size.

The costs associated with the fuel cell stack structure may be broken down based on the baseline fuel cell technology design, high volume with improvements in fuel cell pipeline design, high volume technology improvements in fuel cell design. The flow fields have the highest cost associated with the fuel cell design. The next highest cost is associated with the catalyst and the substrate.

With additional improvements in the packaging design, it is possible to shrink the stack dramatically without changing the electrochemistry of the fuel cells. Design changes being considered and implemented by the stack developers include utilization of thinner cooling plates, smaller humidification sections, use of lightweight clamping materials, and smaller manifolds. With the new designs, there is a five-fold reduction in the weight and almost a three-fold reduction in volume of the fuel cell. Additional weight and volume drops will be 30 to 50% in the next few years.

The production costs analyses indicate that the current cost of fuel cell stacks is in excess of $3,000/kW. There are four major elements associated with the large reduction in cost:

1. The increases in the fuel cell power density described in the previous section will result in lower material requirements (approximately 1.2 kg of material are used per kW of stack output).
2. Costly materials have been reduced in quantity or replaced in quantity or replaced. For instance, platinum catalyst loadings have been reduced from a total of $4 mg/cm^2$ to less than $0.4 mg/cm^2$.
3. Increases in production volume will result in much more cost-effective manufacturing techniques. Low production volume and the complex chemistry of the polymer electrolyte membrane result in the costs, which are 10 to 20 times higher. In addition, several stack manufacturers have developed carbon/plastic structures, which can be molded thereby eliminating costly and time consuming machining operations of fine grain graphite separator plates.
4. Production volume alone can reduce fuel cell stack costs for technology to below $250/kW.

It is a well known fact that large market demands for the fuel cell technology will require a operation of automotive vehicles with conventional fuels (gasoline) as well as alternative fuels including ethanol and methanol. The technology development projects for processing fuels have been significant. The Fuji methanol reformer takes only 45 seconds to respond to the accelerator commands. Significant progress has also been made in developing the required compact fuel processing subsystems including:

- The successful operation of a compact, quick-start steam reformer-based fuel processor using methanol within a complete fuel cell system (General Motors)
- The efficient operation of partial oxidation reformer configurations using multiple fuels including ethanol and gasoline
- Testing of technology options (selective partial oxidation, anode air bleed, and improved anode catalyst) to reduce CO to levels consistent with reliable fuel cell stack operation (Ballard, General Motors)

Cost and packaging analyses of these configurations indicate the potential for achieving fuel processor subsystem costs of under $50/kW.

The long term and reliable operation of fuel processing systems under the full range of automotive driving conditions still needs investigation.

The fuel-cell processing components of the Daimler-Chrysler concept are described in the following discussion. Fuel is heated to convert gasoline from the liquid form into gas in a fuel burner/vaporizer—a canister currently about 150 mm in diameter and 500 mm long. This ensures that there is cleaner, soot free combustion of the fuel.

The vaporized gasoline is processed into vapor using the partial oxidation (POX) process fuel reactor. The POX fuel reactor is a metal canister with a spark plug about 350 mm in diameter and approximately 560 mm in length. By limiting the amount of air in the low-pressure environment, H_2 and CO are produced. Sulphur from the gasoline vapors is converted into H_2S gas and then filtered from the vapor from the conversion point.

Since CO tends to poison the fuel cells, it is important to eliminate or reduce the CO levels to extremely low levels (less than 10 ppm). Water is introduced as steam and acts as a catalyst. Acting with copper oxide and zinc oxide catalysts, the CO converts to CO_2. Additional H_2 fuel is also produced in this stage.

While the water vapor layer has been converted from about 30% CO to a hydrogen rich gas mixture containing only about 1% in the water-gas shift, there are still 10,000 ppm of CO. In the preferential oxidation stage, air is injected through the gas mixture. The air reacts with the remaining CO over the Pt catalyst to produce CO_2, leaving a very small amount of CO (only trace levels at less than 10 ppm).

The unit for the steam and the air-injection process (water-gas shift and preferential oxidation), along with a device for sulphur removal is of the same size as the burner/vaporizer. A small air compressor is approximately 300 mm in diameter and 300 mm length.

The process requires a heat exchanger to maintain effective performance since the clean gas must be cooled to about 80°C—the temperature at which the fuel cell operates at its best. The fuel-cell radiator is comparable to the size of the radiator in the present day car and can handle all waste heat.

The system includes an exhaust for the byproducts: CO_2, N_2, and H_2O—much of which will be reused in the reformation process.

A less critical question for the entire system is how much battery power will be required. Batteries may be needed to heat the system to operating temperatures, energize the spark plugs, and provide propulsion, auxiliary power at the same time. The preceding provides the background for positioning fuel cell technology as a key component of a strategy for achieving the goals of the advanced vehicle initiative within a reasonable balance of risk and benefit, specifically:

- Fuel cell technology has promise of being able to achieve performance/cost goals required of the advanced vehicle development program.
- There are no known technical barriers to verifying that these goals can be demonstrated in automotive hardware in a two- to four-year time frame assuming that the needed levels of support have been provided.
- The potential for PEM based fuel cell to meet transportation requirements is now being recognized in Europe and Japan for developing cars and light trucks. This is a major motivation for the rapidly increasing commitment to related technology developments.

COMPARISON OF FUEL CELL TECHNOLOGIES

Several types of fuel cells are being developed for a large number of commercial applications. The proton-exchange membrane fuel cell is generally considered the most promising fuel cell for cars and light trucks. Its low operating temperatures allow it to provide a quick startup. In addition the power it generates for its weight and size is suitable for light-duty vehicle traction.

One likely concern is that the H_2 fuel stream can be easily contaminated with other gases. The likely contaminant gases include CO, CO_2. Exposure to CO from the input fuel stream reduces the PEM fuel cell's (PEMFC) performance. Power requirements of three different car sizes using the PEMFC are calculated—a small car with a 1.3 ltr engine and a 1,250 kg mass, a mid-size car with a 3 ltr engine and 1,775 kg mass, and a full-size van with a 5 ltr engine and 2,800 kg mass. A vehicle simulation program provides the power requirements using the above weight and capacity and maintains a 80 km/hr 7% grade. The small car needs a 50 kW fuel cell with a mass of 150 kg, the mid-size car needs a 95 kW cell with a mass of 285 kg, and the van requires a 125 kW cell with a mass of 375 kg. An aluminum-intensive mid-size car saves an additional 285 kg and reduces the fuel power requirements to 80 kW. The fuel cell power system mass is based on the gravimetric power density of a current stack of 3 kg/kW, and the volume of the fuel cell power system, expressed in ltrs has the same numeric value as the mass in kg or 3 ltr/kg. In addition, the total package of the fuel-cell system powered vehicle includes the compressed hydrogen gas tank.

A small car capable of a driving range of 560 km, requires 5 kg of H_2, the mid-size car requires 6 kg of H_2, and the van requires 11 kg of H_2. Onboard storage of H_2 in a small size car at a pressure of 34 Mpa requires a storage volume of 150 ltr. A mid-size car requires a storage volume of 180 ltr and the van requires a storage volume of 330 ltr. The volume of the compressed gaseous hydrogen, the fuel cell system and the available volumes of the three vehicles demonstrate that it is difficult to package the fuel cell power system. In addition to the available volume, it is necessary to consider the geometric dimensions and the shape of the components along with the weight distribution of the components. The aluminum mid-size car has a limited cargo space and the hydrogen storage tanks fill the trunk of the vehicle. The van is the only vehicle that can accommodate the fuel cell engine, its accessories, and the compressed gas fuel tank without the loss of cargo space and passenger seats.

The other fuel cell under development is the Phosphoric acid based fuel cell (PAFC). It has a lower power density and is not suitable for light transportation vehicles. The alkaline fuel cell (AFC) has a reasonable power density and operates at low temperatures. However, the performance of this fuel cell can be degraded by CO_2 contamination from the input fuel stream. The solid oxide fuel cells (SOFC) have very high power, but their operating temperature may require long start-up times. PEM, phosphoric acid, and alkaline fuel cells are suited for heavy transportation vehicles including buses, locomotives, and large trucks. A comparison of the fuel cell technology in Table A–2 summarizes the characteristics based on power density and operating temperature.

Another fuel cell that has gained interest is the Polymer electrolyte fuel cell (PEFC). This fuel cell can be broken down into the membrane-electrode (M&E) assembly. The electrode component of the M&E assembly is a thin film, 5 to 50 μm thick layer, containing a dispersed Pt catalyst. The catalyst is in good contact with the ionomeric membrane, the central material serves as the electrolyte and the gas separator in the cell. The ionomeric membrane electrolyte is typically 50 to 175 μm thick.

Table A–2 Comparison of fuel cell technology.

Fuel Cell	Availability	Power Density	Operating Temperature
PEM	Commercial	Medium	Low
PAFC	Commercial	Low	Moderate
AFC	Commercial	Medium	Low to Moderate
SOFC	Prototype	High	High

The M&E assembly of the PEFC consists of an ionomeric membrane with thin catalyst layers bonded onto each of its two major surfaces. The M&E assembly has a generic structure of an electrochemical cell: electrode/electrolyte/electrode, packaged in the form of the sandwich of the three thin films. The gas diffuser (or backing) layers that are in immediate contact with the catalyzed membrane are made of porous carbon paper, or a carbon cloth. These layers are 100 to 300 µm thick and are wet-proofed by treatment with PTFE (poly-tetrafluoroethylene or Teflon). The gas diffuser layers serve as the direct and uniform access to reactant gases, H_2, O_2, and to the catalyst layers. The diffuser layers assist with gaseous phase access without having the gases to actually diffuse through films of liquid water.

With the high-density graphite current collector plates and the machined flow fields for effective distribution of reactant gases along the surfaces of the electrodes, the fuel cell assembly is complete. The PTFE mask confines the gas flow to the active areas of the cell. The PTFE mask together with the catalyzed membrane provides an effective sealing. In the fuel cell stack, many such cells are stacked together to generate the voltage required for a given application.

In the PEFC, the Pt/C powder has to be intermixed carefully with a recast ionomer to provide sufficient ionic conductivity within the catalyst layer. Thus the catalyst layer may be described as the Pt/C/ionomer composite, where each of the three components is uniformly distributed within the volume of the layer.

The ionomeric membrane is a proton conducting polymeric membrane and also forms the most unique element of the polymer fuel cell. The membrane layer most commonly used in the development of the PEFC is the perfluorocarbon sulfonic acid ionomer. DuPont Chemicals, W.L. Gore, and Asahi Chemicals are the commercials manufacturers of this ionomer. These membranes serve as the long-term, stable layers under both oxidative and reductive environments. The high conductivity levels achieved by these layers translate to low surface area resistance values as low as $0.05\,\Omega$-cm^2 for a membrane 50 µm thick. The thin membrane also serves as an effective gas separator. With the permeability of both hydrogen and oxygen through the separator membrane of the order of magnitude $10\,mol/cm$-sec-atm. This is equivalent to current density of $1-10\,mA/cm^2$ through a 100 µm thick membrane in a fuel cell. The leakage current is at or below 1% of the operating current of a PEFC—$1\,A/cm^2$ or higher.

The ionomeric membrane has a limited operating temperature range. The first limitation forces the fuel cell operation to be limited to temperatures below 100°C. The fuel cell may operate at a higher temperature of 120°C, but this will be require additional humidification of the fuel cell membrane. Increased humidification is required in order to maximize the conductivity of the fuel cell. The costs associated with the membrane are significant, but these come down significantly as the market for the membrane increases significantly especially with increased applications in electric vehicle designs.

The gas diffuser is a porous backing layer immediately behind the catalyst layer. The gas diffuser serves as an effective reactant gas supply to the catalyst layer and also an effective water supply and removal system. The Teflon (PTFE) layer must be wet-proofed to ensure that at least part of the pore volume in the cathode backing remains free of liquid water in the operating cell of the fuel cell. The combination of the degree of porosity and the amount of Teflon (PTFE) together determine the efficiency of the gas diffuser layer. The backing material is made of a highly stable and conductive material. To date, most of the PEFC backing materials are made of porous carbon paper or cloth.

The outermost layers on both the sides of the unit cell are the current collector plates. These plates contain the gas flow fields shaped as a continuous single serpentine channel, parallel channel, and series-parallel combination channels. The flow-field geometry design is important for effectively removing water from the cathode. The current collector plate becomes the bipolar plate in the PEFC stack capable of high electron conductivity and is impermeable to oxygen and hydrogen gas. Carbon, stainless steel, and titanium have been considered as potential current collector materials or bipolar plates for the fuel cell design applications.

FUEL-CELL EMISSIONS

Emissions in the case of fuel-cells are their advantage over the IC combustion engines. Gasoline vehicles emit a significant amount of air pollutants, air toxins, and CO_2. Air toxins include benzene, formaldehyde, 1,3-butadiene, and acetaldehyde. Air pollutants are emissions regulated by the Clean Air Act: hydrocarbons (HC), carbon dioxide (CO), nitrogen oxides (NO_x), sulphur oxides (SO_x), and particulate matter (PM). The health effects of these pollutants range from headaches, physiological stress, to seriously sustained respiratory damage.

Although formulated gasoline and preheated catalysts are expected to reduce the pollutant emissions, the magnitude of these reductions can vary significantly because these systems can be tampered with. In the case of the failure of the control systems, the pollution due to gasoline vehicles may increase significantly. In addition, driving conditions, heavy acceleration, or excessive stopping and starting can increase emissions. Hydrogen fuel cell vehicles, on the other hand, are nonpolluting. Tampering, technology failure, and driving conditions cannot cause pollutant or toxic emissions from a hydrogen fuel cell vehicle. The introduction of the fuel cell vehicles greatly simplifies gas emissions regulations and enforcement.

In addition to pollution from the vehicles, production and distribution of the fuels can release significant amounts of air pollutants. Table A–3 shows the relative reduction in emissions compared to a IC engine vehicle and fuels used in the fuel cell and traction battery based electric vehicles. Gasoline production from the oil refineries contributes to the emissions of all pollutants, including gasoline distribution, storage, and refueling greatly increases hydrocarbon pollution. Although battery-powered EVs are zero-emission vehicles, the power plants that produce the electricity to charge the traction batteries may emit pollutants. Hydroelectric plants produce very little air pollution but the coal-fired plants produce very high air pollution. However, renewable sources producing electricity to recharge batteries emit zero emissions. Reformation of natural gas results in production of CO_2 and trace amounts of pollutants. However, hydrogen produced, using wind, bio-mass, or solar energy, has no emissions associated with the production of electricity.

Table A–3 illustrates the percent reduction or increase in pollutants relative to the gasoline vehicles in 2000.

POLICY DIRECTIVES FOR CLEAN AIR ACT

The federal Clean Air Act and its amendments set specific air quality requirements for the most polluted cities in the United States. Some

Table A–3 The percent reduction or increase in pollutants relative to gasoline vehicles in 2000.

Fuel/Vehicle	NMOG	CO	NO_x	SO_x	PM	Greenhouse Gases
BPEV	−95	−99	−56	321	153	−37
H2/FCEV	−100	−100	−100	−100	−100	−65
FCEV or BPEV	−100	−100	−100	−100	−100	−94
Baseline emissions on gasoline (g/km)	0.48	3.81	0.28	0.035	0.01	282.5

states are following California in adopting strict regulations for emissions from IC combustion vehicles—light duty trucks and passenger cars. The dependence on IC engine based vehicles continues to produce staggering detrimental effects on human health. Emissions from automotive applications contribute to 50% of all air pollution regulated by the Clean Air Act. In the large cities across the United States, the total pollution due to transportation exceeds 60%. This includes the toxic air pollutants such as benzene, formaldehyde, CO, and CO_2. The Partnership of New Generation of Vehicles is the only voluntary effort relating to improving fuel economy, which calls for the production of vehicles that will get at least three times the average current fuel economy.

In 1990, California passed the zero-emission vehicle mandate as part of the low-emission program. The mandate requires that 2% of the cars produced for sale in 1998 (approximately 40,000 vehicles) and beyond be zero-emission vehicles. By 2003 there must be 10% (approximately 200,000 vehicles) zero-emission vehicles produced for sale. Zero-emission as defined by the act means that the vehicle emits no exhaust or evaporative emission of any kind. Currently, the EV is the only commercially available vehicle that meets this requirement. Similar regulations are under construction by other states. Fuel cell vehicles with onboard hydrogen will emit no pollutants. The vehicles with onboard methanol use a reformer to produce hydrogen and emit small amounts of CO, nitrogen oxides, hydrocarbons, and some greenhouse gases.

Tailpipe emissions of pollutants from the average automobile has continued to decline since 1968 by more than 60% as a result of tighter standards set in the Clean Air Act. Using the strategies like new fuel reformulation, engine controls, and tailpipe pollution control may be less effective than the use of emerging alternative fuels and renewable-energy based propulsion systems.

As part of the development program, the federal government provides tax credits for the purchase of fuel-cell vehicles. Credits are issued for purchasing the zero-emission vehicles that exceed the emissions reductions required by the Clean Air Act. These tax credits can be traded and will in turn be incentives to purchase EVs and fuel cell-based vehicles.

RECOMMENDATIONS

The transition to alternate, renewable-fuels and hydrogen vehicles face hurdles to success and intense opposition from the automotive and oil industries. The following may be used as recommendations towards developing a market entry strategy:

- Build public awareness and confidence in alternate-fuel technology vehicles.
- Develop consumer marketing of alternate-fuel technology vehicles through demonstrations of the urban bus EVs. Explain the advantages of the alternate-fuel technology vehicles.
- Involve the local and federal governments to promote the alternate-fuel vehicle development and sales.
- Support educational, private sector, and government research programs for the development and marketing of alternate-fuel vehicles.
- Develop a national alternate-fuel vehicle transportation program to promote renewable energy programs at the state level.

Although the direct hydrogen filled PEMFC engine-based vehicle is a true zero-emission vehicle, there are several challenges that need to be overcome. Efforts on the part of fuel cell developers and the automotive companies will be essential to make the production and the commercialization of the fuel cell based vehicle successful.

Appendix B VEHICLE BATTERY CHARGING CHECKLIST/LOG

Date	Time of Charge	Dis/Connect	Battery Temp.	SOC	Comments	KWhr Meter

Appendix C DAY 1/2/3 RANGE AND CHARGE TEST LOG

Project No:	
Vehicle Identification:	
Test Date:	
Test Engineer:	
Test Driver:	

Left Front Tire Pressure:	Left Rear Tire Pressure:
Right Front Tire Pressure:	Right Rear Tire Pressure:
Total Vehicle Weight:	

Test Track Location:	Test Track Gradient:
Ambient Temperature:	
Initial Track Temperature:	Final Track Temperature:
Initial Wind Direction:	Final Wind Direction:
Initial Wind Velocity:	Final Wind Velocity:

Appendix D SPEEDOMETER CALIBRATION TEST DATA LOG

Project No:	
Vehicle Identification:	
Test Date:	
Test Engineer:	
Test Driver:	

Initial Odometer Reading:	Final Odometer Reading:
Initial SOC:	Final SOC:
Total Vehicle Weight:	

Data Acquisition Display	Vehicle Speedometer
0 mph	
5 mph	
10 mph	
15 mph	
20 mph	
25 mph	
30 mph	
35 mph	
40 mph	
45 mph	
50 mph	
55 mph	
60 mph	
70 mph	
80 mph	

Appendix E ELECTRIC VEHICLE PERFORMANCE TEST SUMMARY

Based on the Electric Transportation Division NiMH battery and conductive performance tests for urban driving and highway (freeway) driving, the following tables summarize the vehicle range tests performed by California Edison.

UR1 Urban Range Test with minimum Payload and no Auxiliary loads

UR2 Urban Range Test with minimum Payload, HVAC on high, Headlights on Low, and Radio on

UR3 Urban Range Test with maximum Payload and no Auxiliary loads

UR4 Urban Range Test with maximum Payload, HVAC on high, Headlights on Low, and Radio on

FW1 Freeway Range Test with minimum Payload and no Auxiliary load

FW2 Freeway Range Test with minimum Payload, HVAC on High, Headlights on Low, and Radio on

FW3 Freeway Range Test with maximum Payload and no Auxiliary load

FW4 Freeway Range Test with maximum Payload, HVAC on High, Headlights on Low, and Radio on

1998 Ford Ranger with NiMH Urban Driving Test

Test	UR1	UR2	UR3	UR4
Payload (lbs)	190	190	1,220	1,220
KWhr Charge	31.74	31.76	32.96	32.95
AC kWhr/mile	0.391	0.431	0.435	0.503
Range (miles)	80.6	73.2	75.3	62.7
Ambient Temp.	63°F	63°F	66°F	69°F

1999 Chrysler EPIC with NiMH Urban Driving Test

Test	UR1	UR2	UR3	UR4
Payload (lbs)	160	160	930	930
KWhr Charge	53.91	50.03	53.02	52.61
AC kWhr/mile	0.663	0.734	0.675	0.823
Range (miles)	82.0	87.8	77.6	63.6
Ambient Temp.	75°F	80°F	79°F	85°F

1999 Toyota RAV4 NiMH Urban Driving Test

Test	UR1	UR2	UR3	UR4
Payload (lbs)	160	160	766	766
KWhr Charge	31.60	33.96	32.72	32.22
AC kWhr/mile	0.329	0.394	0.360	0.434
Range (miles)	92.8	84.8	89.5	68.9
Ambient Temp.	68.5°F	75.3°F	80°F	87°F

1998 Ford Ranger with NiMH Freeway Driving Test

Test	FW1	FW2	FW3	FW4
Payload (lbs)	150	150	1190	1190
kWhr Charge	31.99	35.17	36.85	33.89
AC kWhr/mile	0.409	0.458	0.480	0.490
Range (miles)	76.5	71.4	74.8	68.8
Ambient Temp.	62°F	66°F	74°F	52°F

1999 Chrysler EPIC Freeway Driving Test

Test	FW1	FW2	FW3	FW4
Payload (lbs)	160	160	930	930
kWhr Charge	54.08	51.54	50.42	55.52
AC kWhr/mile	0.542	0.674	0.596	0.799
Range (miles)	99.3	75.3	80.3	68.6
Ambient Temp.	86°F	88°F	83°F	101°F

1999 Toyota RAV4 NiMH Freeway Driving Test

Test	FW1	FW2	FW3	FW4
Payload (lbs)	180	160	766	766
KWhr Charge	32.54	31.33	31.79	31.88
AC kWhr/mile	0.404	0.406	0.428	0.418
Range (miles)	79.9	76.6	76.2	75.3
Ambient Temp.	89.5°F	88°F	79°F	81°F

BIBLIOGRAPHY

Anderson, I. E., et al., 1998, Benefits of rapid solidification processing of modified LaNi$_5$ alloys by high pressure gas atomization for battery applications, *Proceedings*, Materials Research Society Symposium, v. 496.

Battery Charging, Electric Transportation Applications, 1999.

Battery mission: To charge and to protect, *Portable Design*, October 1997.

Battery pack safety and handling procedures, Unpublished Document, 1996.

Battery state of charge, application, and prediction, Johnson Controls R&D Document, 1996.

Bell, R. A., December 1996, Cold-weather impacts on electric vehicles, North American EV and Infrastructure Conference.

Bernardi, D., and M. K. Carpenter, August 1995, Mathematical model of the oxygen recombination lead-acid cell, *Journal of the Electrochemical Society*, v. 142.

Birch, P. K., 1984, Thermal management of electric vehicle battery system, *Proceedings*, Conference of Energy Conservation in Industry.

Buchmann, I., June 2000, Give your battery the smarts it needs, *Portable Design*.

Cannon, J. S., April 1997, Clean hydrogen transportation: A market opportunity for renewable energy, Renewable Energy Policy Project, Issue Brief.

Carl, B., Temperature termination and the thermal characteristics of NiCd and NiMH batteries, Falcon, *Proceedings of the 1994 IEEE Wescon Conference*, Anaheim, California.

Casacca, M. A., et al., September 1992, Determination of lead-acid battery capacity via mathematical modeling techniques, *IEEE Transactions on Energy Conversion*.

Charge methods for nickel-metal hydride batteries, August 1998, *Ni-MH Batteries Handbook*, Panasonic.

Charging station for electric vehicles, September 1992, *Proceedings*, 11th IEVS.

Cole, G., A generic SFUDS battery test cycle for electric vehicle batteries, Idaho National Engineering Laboratory.

Commercialization of nickel-metal-hydride batteries for electric and hybrid vehicles, April 1999, Office of Transportation Technologies.

Commercialization of Nickel-Metal-Hydride Batteries for Electric and Hybrid Vehicles, brochure, April 1999, Office of Transportation Technologies.

Craven, W. A., 1996, *Horizon Sealed Lead Acid Battery in Electric Vehicle Application*, Electrosource, Inc.

Davies, E. D., April 1996, Suppressing the growth of dendrites in secondary Li cells, *NASA Tech Briefs*.

Deterioration estimation method for 200 Ahr sealed lead-acid batteries, 1994, *NTT R&D*.

Dhameja, S., and Min Sway-Tin, February 1997, *Traction Batteries—Their Effects on Electric Vehicle Performance*, SAE International Congress and Exposition.

Dick. B., et al., July 1998, A battery of analysis, *Telephony*.

Dietz, H., et al., January 1991, On the characteristics of oxygen recombination in sealed lead-acid batteries, *Journal of Applied Electrochemistry*, v. 21.

Djong-Gie Oei, February 1997, Fuel cell engines for vehicles, *Automotive Engineering*.

Dunbar, J., 1994, High-performance nickel metal hydride batteries, *IEEE Proceedings*, Wescon Conference.

Electric Vehicle Operation and Range Calculation Using Pb-acid Batteries, 1995, unpublished.

Electrolyte Spillage and Electric Shock Hazards, 1994, National Highway Traffic Safety Administration (NHSTA) and Department of Transportation (DOT) Request for Comments.

Energy conversion devices, 1996, *Product Development and Commercialization*, brochure.

Fast charging advances the art of refuelling electric vehicles, May 1991, *Proceedings*, 24th ISATA.

Fetchenko, M. A., et al., 1992, Selection of metal hydride alloys for electrochemical applications, *Proceedings of the Electrochemical Society*, v. 92(5).

Fetcenko, M. A., et al., 1988, *Hydrogen Storage Materials for Use in Rechargeable Ni-Metal Hydride Batteries*, 16th International Power Sources Symposium.

Fuel cell technology for transportation applications: Status and prospects, June 1995, *Executive Summary*, report.

Furukawa, N., September 1994, Development and commercialization of nickel-metal secondary batteries, *Journal of Power Sources*, v. 51.

Gottesfeld, S., 1999, *The Polymer Electrolyte Fuel Cell: Materials Issues in a Hydrogen Fueled Power Source*, Argonne National Laboratories.

Gottwald, T., Z. Ye, and T. Stuart, 1994, *Equalization of EV and HEV Batteries with a Ramp Converter*, University of Toledo.

Gu., H., et al., December 1987, Mathematical model of a lead-acid cell, discharge, rest and charge, *Journal of Electrochemical Society*.

Hartmannn, M. J., 1994, Battery terminal voltage calculations, *Proceedings of the American Power Conference*.

Horie, H., et al., 1994, Development of battery simulator for EV, technical notes, *JSAE Review*.

Hullmeine, U., et al., 1989, Effect of previous charge/discharge history on the capacity of the $PbO_2/PbSO_4$ electrode: The hysteresis or memory effect, *Journal of Power Sources*, v. 25.

Hydrogen Fuel Cell Vehicles, February 1995, Union of Concerned Scientists.

IEEE Recommended Practice for Maintenance, Testing, and Replacement of VRLA Batteries for Stationary Applications, April 1994, IEEE.

Ikoma, M., S. Yuasa, et al., 1998, Charge characteristics of sealed-type nickel/metal hydride battery, *Journal of Alloys and Compounds*, v. 267.

Jay, B. E., et al., 1995, *Performance of the Horizon Advanced Lead-Acid Battery*, Electrosource, Inc.

Jost, K., February 1997, Gasoline-reforming fuel cell, *Automotive Engineering*.

Keller, S. A., et al., September 1991, Thermal characteristics of electric vehicle batteries, *SAE Proceedings*.

Kiessling, R., 1994, A battery model for monitoring and corrective action on lead-acid EV batteries, *IEEE Proceedings*.

Koehler, U., et. al., September 1998, Nickel-metal-hydride and lithium-ion batteries for automotive applications, *Special Report*, VARTA.

LaNi$_{5-x}$M$_x$ Alloys for Ni/Metal Hydride Electrochemical Cells, May 1999, NASA Jet Propulsion Laboratory, California.

Lee, J., et al., 1982, Three-dimensional thermal modeling of EV batteries, *Electrochemical Society Extended Abstracts*.

Level III Charging of Electric Vehicles, Electric Transportation Applications, 1999.

Magnuson, D. C., et al., September 1994, *Status and Prospects of Metal Hydrides for Nickel-Metal Hydride Secondary Batteries, Materials Letters*.

Mason, W. T., et al., 1994, Hybrid EVs versus pure EVs: Which gives greater benefits, *SAE Proceedings*.

Masserant, B. J., and T. A. Stuart, October 1994, *A High-Frequency DC/DC Converter for Electric Vehicles*, IEEE Workshop on Power Electronics in Transportation.

Masserant, B. J., and T. A. Stuart, 1994, *A Maximum Power Transfer Battery Charger for Electric Vehicles*, internal report, University of Toledo, Ohio.

Mathematical modeling of a nickel metal-hydride cell, January 1995, *Proceedings*, AIP Conference, v. 325 (1).

Misra, S., et al., 1994, *AC Impedance/Conductance Testing of VRLA Batteries*, C&D Charter Power Systems, Inc.

Mukerjee, S., et al., September 1997, Effect of Zn additives to the electrolyte on the corrosion and cycle life of some AB$_5$H$_x$ metal hydride electrodes, *Journal of the Electrochemical Society*, v. 144.

Murray, C., November 1991, Battery aims to solve electric vehicle woes, *Design News*.

Nagasubramanian, G., et al., 2000, *Electrochemical Characteristics of Lithium-Ion Cells*, Sandia National Laboratories.

Nor, J. K., September 1992, Charging station for electric vehicles, *Proceedings, 11th IEVS*.

Nor, J. K., May 1991, Fast charging advances—The art of refuelling electric vehicles, *Proceedings*, 24th ISATA.

Ohta, K., et al., 1994, Nickel hydroxide electrodes: Improvement of charge efficiency at high temperature, *Proceedings*, The Electrochemical Society, Inc., v. 94.

Ovshinsky, S. R., et al., April 1993, A nickel metal hydride battery for electric vehicles, *Science*.

Ovshinsky, S. R., et al., May 1994, Ovonic NiMH batteries for portable and EV application, *Proceedings*, 11th International Seminar on Primary and Secondary Battery Technology and Application.

Ovshinsky, S. R., et al., May 1991, Ovonic Ni-Metal hydride batteries for electric vehicles, *Proceedings*, 24th ISATA Symposium.

Owen, F., March 1999, *Battery Protection*, Raychem Corporation, PCIM.

Pavlov, D., and R. Popova, 1970, Mechanism of Passivation Processes of the Lead Sulphate Electrode, *Electrochimica Acta*.

Ponticel, P., June 1997, Driving for cleaner cars, *Automotive Engineering*.

Protogeropoulos, C., et al., 1994, Battery state of voltage modeling and algorithm describing dynamic conditions for long-term storage simulation in a renewable system, *Journal of Solar Energy Science and Engineering*.

Prout, L., August 1993, Aspects of lead/acid battery technology designing for capacity, *Journal of Power Sources*, v. 46.

Rechargeable Nickel-Metal Hydride Application Manual, 1998, Energizer/Eveready Corporation.

Reid, M., February 1997, The never-ending quest for battery life, *Portable Design*.

Reisner, M. E., and M. Klein, December 1994, Low-cost plastic-bonded bipolar nickel-metal hydride EV battery, *Proceedings*, EVCS-12, Anaheim, California.

Reisner, D. E., and M. Klein, June 6–9, 1994, Sealed bipolar nickel-metal hydride battery, *Proceedings*, 36th Power Sources Conference.

Reisner, D. E., and M. Klein, March 4–7, 1996, Stackable wafer cell-type bipolar alkaline battery: milestones and applications, *Proceedings*, 13th International Seminar on Primary and Secondary Battery Technology and Applications.

Reisner, D. E., M. Klein, et al., November 11–15, 1995, Low-cost plastic bonded bipolar Ni-MH EV battery, *Proceedings S/EV 95*.

Rudenko, M. G., August 1993, Comparison of the discharge of lead-acid battery positives and negatives, *Electrokhimia*.

The secret life of cells, April 1998, *Portable Design*.

Sharpe, T. F., and R. S. Conell, 1987, Low-temperature charging behaviour of lead-acid cells, *Journal of Applied Electrochemistry*, v. 17(4).

Soileau, R. D., January–February 1994, Diagnostic testing program for large lead acid storage battery banks, *IEEE Transactions on Industry Applications*.

The state of lithium rechargeables, March 1998, Designers Notebook, *Portable Design*.

Stuart, T. A., 1994, *Battery Management Research for Electric Vehicles*, unpublished technical paper, University of Toledo.

Testing and Development of Electric Vehicle Batteries for EPRI Electric Vehicle Transportation Program, 1985, technical paper, Argonne National Laboratory.

The Transportation Program, Battery Development for Electric Vehicles, 1994, EPRI.

The Transportation Program, Inductive Charging for the Electric Vehicle Catalog, 1993, Electric Power Research Institute.

Tung, S. T., and D. C. Hopkins, et al., February 1993, Extension of battery life via charge equalization, *IEEE Transactions on Industrial Electronics*, v. 40.

USABC Electric Vehicle Battery Test Procedures Manual, Revision 1, July 1994.

Valeriote, E. M., T. G. Chang, et al., January 1994, Fast charging of lead acid batteries, *Proceedings*, 9th Annual Battery Conference on Applications and Advances.

Valeriote, E. M., and D. M. Jochim, 1992, Very fast charging of low resistance lead-acid batteries, *Journal of Power Sources*, v. 40.

Valeriote, E. M., et al., April 1991, Very fast charging of lead-acid batteries, *Proceedings*, 5th ILZRO.

Valoen, L. O., S. Sunde, and R. Tunold, *Characterization of the Electrochemical Properties of Metal Hydrides by AC Impedance*.

Valoen, L. O., et al., August 1996, An impedance model for electrode processes in metal hydride electrodes, *Conference Proceedings*, MH96.

Venkatesan, S., et al., 1989, Development of ovonic rechargeable metal hydride batteries, *IEEE Conference Proceedings*.

Venkatesan, S., et al., 1989, Polarization studies with ovonic metal hydride batteries: Part I, *IEEE Proceedings*.

Vutetakis, D. G., et al., June 1992, Effect of charge rate and depth of discharge on the cycle life of sealed lead-acid aircraft batteries, *Proceedings*, 35th International Power Sources Symposium.

Wilson, H. D., New SAE standards for battery charge acceptance, *IEEE Proceedings*.

1998 Ford Ranger Electric Vehicle Specifications.

1998 Ford Ranger EV Catalog.

1999 Electric Vehicle (NiMH Batteries) Performance Characterization Summary, Southern California Edison.

INDEX

AB_2/AB_5 alloys, 9
Absorbed glass mat (AGM) batteries, 7–8, 34, 98; *See also* Valve regulated lead-acid batteries
Accelerated reliability testing, 167–71
Acceleration, 2, 59, 63–64, 66, 182–83
Accidents, 149–50, 154, 167, 189–90
AC conductance tests, 92–93
Acid spills, 151
AC impedance tests, 92–93
AC motors, 3
Active equalization, 141–42
Adhesion of active paste, 31
Aerodynamic drag, 65, 131, 182
AGM (absorbed glass mat) batteries, 7–8, 34, 98; *See also* Valve regulated lead-acid batteries
Air emissions, 187, 191, 193, 201–2
Air-flow models, 138–40
Alkaline fuel cells (AFC), 199
Alternating current (AC) motors, 3
Ambient temperature model, 139
Ampacity, 52
ANSI/IEEE 450 standard, 53, 60–64
Antimony, in Pb-acid batteries, 6, 34–35
Arrenhius equation, 179
Auxiliary power units (APUs), 68, 186–87
Auxiliary systems, 170, 185

Batteries; *See also names of individual battery types*
  choice of, 18–21
  definition and components, 4
  power calculation, 184–85
  safety design, 150–53, 188–90
  smart, 147–48
  USABC on, 4–5
  12 V auxiliary, 148–49
Battery acceptance test, 60–62
Battery condition indicators, 146

Battery degradation, fast charging and, 108–10; *See also* Cycle life
Battery modules, 4, 140–41; *See also* Battery packs
Battery monitors (BMONs), 88, 103, 127, 136, 147–48, 155
Battery packs
  capacity determination, 118–19
  charging/discharging patterns, 117–19, 128–30, 136
  cold temperature, 131–32
  components monitored during testing, 169
  design, 41–42, 133, 140–41
  insulation breakdown detection, 157
  nonuniform temperature, 117, 177
  state of charge calculations, 48
  thermal management, 137–41
  voltage calculation, 176
  voltage cut-off point, 118
  weak cells, 61, 126, 133–34
Battery Performance Management System (BPMS), 133–48
  charge indicators, 144–46
  charge protectors, 142
  charging control, 141–48
  charging/discharging monitoring, 134–35
  components, 134
  design analysis, 140–41
  diagnostics control, 147–48
  model of, 135–36
  thermal management, 137–41, 152
  thermistors, 142–45
  typical configuration, 136–37
Battery performance test, 60–62
Battery scaling, 188
Battery service capacity test, 60
Battery testing, 161–90; *See also* Electric vehicle testing
  capacity discharge, 51–53
  charge completion on oxidation, 166

conductance, 92–93
constant current discharge, 164
constant power, 164
core battery performance, 163–66
crash tests, 189
cycle life, 171–73
fast charge, 166
NiMH modeling, 174–76
partial discharge, 165
peak power, 164
performance/acceptance, 60–62
recommendations, 173–76
service capacity, 60
standloss, 165
testing approach, 161–62
thermo-electrochemical model, 176–88
variable power discharge, 115, 164
vibrations, 166
Battery vibration test, 166
BMONs (battery monitors), 88, 103, 127,
    136, 147–48, 155
Boost charges; See Equalization charges
Bottom-pour casting, 32
BPMS; See Battery Performance Management
    System
Braking, regenerative, 2–3, 24, 68, 134, 183
Braking system component monitoring,
    170–71
Breakdowns, vehicle, 168
Building codes, 90, 159

Calcium, in Pb-acid batteries, 6, 34–35
Capacity, battery, 43–68
    ampacity, 52
    battery acceptance test, 60–62
    Battery Performance Management System,
        135–36
    battery performance test, 60–62
    battery service capacity test, 60
    calculation, 187–88
    charge rate vs., 76, 96, 106–7
    C ratings, 54–57, 119–20
    definitions, 54–57
    during discharge, 51–53, 119–20, 123–27
    fuel gauges, 145–46
    grid corrosion and, 31
    Li-ion batteries, 58–59
    NiMH batteries, 39, 54–57, 123–27
    positive vs. negative electrodes, 82
    recovery, 53–54
    360-second discharge test profiles, 62–63
    state of charge regulation and, 49
    sulfation, 35, 52

temperature dependence, 44–46, 118–19,
    135, 177
variable power discharge test, 62–63
VRLA batteries, 117–19
weak cells, 61, 126
Capacity discharge testing, 51–53
Carbonate, molten, 16
Carbon compounds, in Li-ion batteries, 10
Carbon monoxide, in fuel cells, 197
Casings, 14, 31
Catalysts
    benefits of, 33–34
    platinum, 15, 17, 192, 195, 197
Catalytic converters, 187
Cathodes, 10, 13, 26, 30; See also Electrodes
C1(T) criterion, 83–84
Cell polarity reversal profile, 125–26
Cells, number of, 41
Cell voltage, 120
Charge acceptance rates
    during fast charging, 96–97, 106–7
    fuel gauges and, 145–46
    heat dissipation, 180
    inefficiencies, 81
    modeling, 175
Charge completion on oxidation test, 166
Charge indicators, 144–46
Charge switching method, 79–80
Charging, battery, 69–94; See also Charging
    stations; Fast charging; Overcharging
    during accelerated reliability testing, 168
    Battery Performance Management
        Systems, 134, 141–48
    charge acceptance rates, 81, 96–97, 106–7,
        145–46, 175, 180
    charge completion on oxidation test, 166
    charge protectors, 142–44
    checklist/log, 205
    components monitored during testing,
        169
    constant current, 172–73, 179–81
    constant current-constant voltage, 69–71,
        86, 89, 98–99, 172
    couplers, 89, 111–12, 151–52, 156
    cycle life and, 172
    data storage on, 136–37
    depolarization enhancement of, 146–47
    efficiency calculations, 94
    environmental influences on, 80–81
    equalization, 32–33, 71, 93–94, 105–8,
        118, 141–42
    excessive, 35–36
    inadequate, 36
    inductive, 111–12

inflection point detection, 87
intelligent chargers, 85–87
Li-ion batteries, 12, 39–40
maximum power, 50
NiMH batteries, 74, 78, 81–87
overview, 4, 155
rate terminology, 95
safety considerations, 90–91
shunts, 141
smart batteries, 147–48
standard receptacles, 103
state of charge calculations, 47–48
temperature-based termination methods, 78–80, 83–85
temperature compensation, 34–35, 70–73
temperature sensing, 74–78
trickle, 86, 89, 96, 109–10
ventilation, 90–91
VRLA batteries, 34–35, 69–71
Charging stations
  charger controls, 102–4, 148
  couplers, 89, 111–12, 151–52, 156
  fast charging, 103–5
  inductive charging, 111–12
  power levels, 88–89, 155–56
  prerequisites, 104–5
  required equipment, 87–88
Circulating-liquid thermal management system, 140
Clamp voltage, during charging, 69–71, 73
Clean Air Act, 202–3
Cobalt oxides, 10–11, 13
Cold weather; See also Temperature
  discharge capacity, 123–24, 130–32
  driving range, 131–32
  fast charging, 98
  fuel gauges, 146
  performance tests, 130–31
  state of charge calculations, 48
Collectors, copper/aluminum foil, 14
Compressed natural gas (CNG), 17–18
Conductance tests, 92–93
Conductive coupling, 89, 156
Constant current charge method, 172–73, 179–81
Constant current-constant voltage (CI-CV), 69–71, 86, 89, 98–99, 172
Constant current discharge test, 164
Constant power test, 164
Containment systems, 151
Coolants, 25, 140
Cooling of batteries, 77, 92, 172–73
Copper oxide catalysts, 197
Corrosivity, 151

Costs
  battery amortization, 171
  battery maintenance, 8
  fuel cell stacks, 192, 195–96
  hydrogen research, 19
  internal combustion engine powertrain, 17
  Li-ion battery production, 20–21
  NiMH production, 21
  operating, 171
  platinum catalyst, 17, 192
  repair, 168
Couplers, charge, 89, 111–12, 151–52, 156
Crashes, 149–50, 154, 167, 189–90
C ratings, 54, 119–20
Current, battery, 49–50, 94
Current density, 25–35
Cycle life
  charge method and, 172
  definition, 53
  depth of discharge, 53–54
  Li-ion batteries, 19–20
  NiMH batteries, 38–39
  peak power demand and, 171–72
  rest periods and, 171–73
  VRLA specifications, 44

Daimler-Chrysler, 5, 18, 197, 212
DC (direct current) motors, 3
Deceleration, 149, 183–84
Dendrite formation, 11, 40, 109–10
Department of Energy (DOE), 193
Depolarization, charge enhancement, 146–47
Depth of discharge (DOD)
  battery resistance, 58
  charge capacity and, 109
  cycle life, 53–54
  discharge rate and, 56–57
  modeling, 175, 188
  temperature dependence, 46
Diagnostics control, 147–48
Diffusion coefficient calculation, 179
Direct current (DC) motors, 3
Discharge current, 43–44, 45
Discharging, 115–32
  Battery Performance Management System, 134–35
  capacity during, 51–53, 119–20, 123–27
  capacity ratings and, 54
  cold weather impacts, 123–24, 130–32
  data storage on, 136–37
  discharge tests, 51–53, 115–17, 164–66

end-of-discharge voltage, 126–27
Li-ion batteries, 12, 127–29
load voltages *vs.* capacity, 119–20
NiMH batteries, 56–58, 121–27, 175
partial, 109, 165
power calculation, 41, 50–51
pulses, 124, 128–29
temperature and, 58, 115–17, 122
termination of, 124–27
voltage profiles, 55–58, 121–26
Disconnects, high-voltage, 149–50
DOD; *See* Depth of discharge
Downtime, vehicle, 168
Drag losses, 65–68, 131
Driveshaft power, 181–83
Driveshaft torque, 182–83
Drive train efficiency, 67, 170
Driving conditions
battery discharge, 115–17, 128–29
change in resistance, 65
depth of discharge, 58
heat calculations, 138
during reliability testing, 167–71
state of charge during, 47
twenty-step test profile, 63
wet, 131
Driving range, 24, 113–14, 130–32, 211–13;
See also Efficiency, battery
Durability tests, 157
Dynamic Stress Test (DST), 115–17, 165
Dynamometer tests, 162

Efficiency, battery, 23–42
charging, 94
elevated temperatures, 28–29, 35
EV body and frame, 24–25
factors affecting, 24
failure modes of VRLA batteries, 31–35
fuel cells, 16–17, 194
NiMH battery formation, 26–31
regenerative braking, 24
Efficiency, vehicle, 66–67
Electrical safety, 151–52
Electrical utilities, fast charging and, 110–11
Electric bus isolation, 149
Electric motors, 3
Electric shock, 153–54, 157, 189
Electric vehicle batteries; *See* Batteries;
names of individual battery types
Electric vehicles (EVs)
air emissions, 202
body and frame materials, 24–25

charging times, 102, 112
components, 3
electronic drive systems, 3
energy balances, 64–68
fast charging range testing, 113–14
fluids, lubricants, and coolants, 25
headlights and taillights, 131
heaters, 131
lubricants, 130–31
motor power, 183
need for, 2, 4–5
operation, 2–3
overall charging efficiency, 94
performance models, 181
performance test summary, 211–13
reliability/durability tests, 157, 167–71
speedometer calibration, 114, 209
tires, 130–31
Electric vehicle supply equipment (EVSE), 88
Electric vehicle testing
accelerated reliability, 167–71
cold weather performance, 130–31
crash, 189
driving range, 113–14
Dynamic Stress Test, 115–17, 165
endurance, 158
extended life, 158
freeway driving, 212–13
operating life, 158
performance safety and abuse, 167
range and charge test log, 207
reliability/durability, 157, 167–71
special performance, 165
specific model results, 211–12
standloss, 165
sustained hill-climb power, 165–66
thermal performance, 166
urban driving, 62, 211–12
Electrodes
cathodes, 10, 13, 26, 30
membrane-electrode-assembly, 195
memory effects, 147
oxidation, 27–28
passivation layer, 53–54
reactions at, 174
surface etching, 30
weight of, 42
Electrolysis
during charging phase, 7, 74, 108–10
with complete discharge, 124–25
in fuel cells, 15
gas buildup, 109
in VRLA batteries, 33, 108–10

Electrolytes
  in absorbed glass mat batteries, 7
  activity variations, 98
  additives, 153
  boiling points, 151
  during charging phase, 7
  conductivity, 59, 92, 179
  corrosive, 151
  forced circulation, 174
  in fuel cells, 16
  leakage during storage, 38
  overcharging and, 11
  over-discharge, 124–25
  polymers, 10–14, 16, 195, 199
  safety design, 150–51, 153–54
  spillage during crashes, 189–90
  sulphuric acid, 6
  temperature and, 35, 162
  types of, 4
Electronic control module (ECM), 2–3, 67,
    127
Electronic drive systems, 3
Emissions, 187, 191, 193, 201–2
End-of-discharge voltage (EODV), 126–27
Energy consumption calculations, 64–68
Energy densities, 12–13, 20, 42, 44
Energy equations, 188
Engine efficiency, 67–68
Entropy changes, 97
Environmental concerns
  air emissions, 187, 191, 193, 201–2
  Li-ion vs. NiMH batteries, 11
  manganese vs. cobalt or nickel, 59
EPIC performance summary, 212
Equalization charges
  active, 141–42
  battery packs, 93–94
  in fast charging, 105–8
  magnitude, 32–33
  to prevent over-discharges, 118
  single VRLA batteries, 71
Etch treatments, 30
Ethanol, in fuel cells, 193, 196
Europe, fuel cells in, 194–95
EVs; See Electric vehicles
Extended life tests, 158

Failure modes, battery
  battery storage conditions, 37
  excessive charging, 35–36
  inadequate charging, 36
  overview, 33
  VRLA batteries, 31–35

Fast charging, 95–114
  battery degradation, 108–10
  charge acceptance ability, 96–97
  charger configuration, 101–5
  constant current-constant voltage
    method, 98–99
  dendrite formation, 109–10
  electrical utilities and, 110–11
  equalization charges, 105–8
  fast charge test, 166
  feedback control, 100
  heat production, 97–98, 101
  inductive, 111–12
  limitations of, 105–6
  maximum voltage-maximum current
    profiles, 101–3
  NiMH batteries, 84
  overcharging, 96–97, 107–9
  prerequisites, 104–5
  process overview, 95–98
  range testing, 113–14
  strategies, 98–101
  temperature and, 97, 100, 108–9
  ultra-fast, 86
  USABC goal for, 166
  voltage/current profiles, 99–100
Federal Urban Driving Schedule (FUDS),
    115, 123, 164–65
Field emission transistors (FETs), 142
Flame arrestors, 7–8
Float charge, 34
Flooded batteries, 6, 129–30, 151, 173
Fluids, EV, 25
Ford Motor Company, 5, 211–12
Ford Ranger test summary, 211–12
Fossil fuel use, 1–2
Freeway driving tests, 212–13
FUDS (Federal Urban Driving Schedule),
    115, 123, 164–65
Fuel cell technology, 191–204
  advantages of, 191–92, 198
  air emissions, 191, 193, 201–2
  alkaline, 199
  batteries, 197–98
  catalysts, 200
  comparison of technologies, 198–201
  cost reductions, 17, 192, 195–96
  Daimler-Chrysler concept, 18, 197
  DOE bus, 193
  efficiency ratings, 194
  fuel cell stacks, 192, 195
  fuel economy, 193
  ionomeric membranes, 200
  market entry strategies, 203–4

molten carbonate, 16
next generation objectives, 193
overview, 14–18
phosphoric acid based, 16, 199
polymer electrolyte, 16, 195, 199–201
proton exchange membrane fuel cells, 15,
    17, 194–95, 198–99
recent developments, 192–94
solid oxide, 16, 199
Fuel gauges, 144–46
Fuel stack technology, 192
Fuel vaporizers, 18

Gas diffusers, 200–201
Gasoline, in fuel cell technology, 196–97
Gel technology batteries, 7–8
General Motors, 5, 196
Glass mat batteries, 7–8, 34, 98
Graphite, 10, 40, 200
Graves, Sir William, 14–15
Gravitational power, 182
Grid corrosion, 31–32
Grid growth, 31–32
Ground-fault circuit interrupter devices, 90,
    149, 152

Hall effect sensors, 136
Headlights, 131
Heat capacity, 75–76, 101
Heat dissipation, 180–81, 197
Heaters, 131
Heat generation models, 138–39, 177–81
Heating, external, 77
Heat pumps, 25
Heat transfer coefficients, 178–81
Heat transfer models, 138–39, 178–81
High-pressure gas atomization, 10
High-speed data bus (HSDB), 136–37,
    147–48
High-voltage wiring systems, 148–50
Hissing, during fast charging, 98
Honda Motor Company, 5
HVAC system, 132
Hybrid vehicles, 68, 127, 186–87
Hydride electrodes, 9
Hydrogen
    combustible levels, 91, 159
    in fuel cells, 15–17, 191
    in NiMH batteries, 26, 39–40, 124–25
    production from gasoline, 18
    storage of, 17
    vehicle requirements, 198–99

ventilation requirements, 90–92, 129–30,
    159
in VRLA batteries, 30–31, 33
weak cells, 133

Impacts, 149–50, 154, 167, 189–90
Impedance tests, 92–93
Inductive charging, 111–12
Inductive coupling, 89, 112, 156
Inertia power, 182
Inertia switch disconnects, 149
Infrastructure Working Council (IWC), 90,
    159
Installation resistance, 91–92
Insulation breakdown detection, 157
Intelligent chargers, 85–87
Internal combustion engines, 68, 201
Inverter/system controller, 184
Ionic conductivity, 179

Japan, fuel cells in, 194–95
Joule's law heating, 180

Knee, of discharge curve, 56

LaNi$_5$ alloys, 9
Lead-acid batteries; See also Valve regulated
        lead-acid batteries
    absorbed glass mat, 7–8, 34, 98
    crash tests, 189
    flooded, 6, 129–30, 151, 173
    formation process, 23
    gel technology, 7–8
    maintenance costs, 8
    overview, 6–8
Lead oxide, 6
Lead sulphate, 6, 28, 54
Li-ion batteries
    advantages, 19
    capacity, 58–59
    cathode materials, 10–11
    cell shape, 127–28
    characteristics, 20
    charging, 12, 39–40
    cobalt oxides, 10–11, 13
    cycle life, 19–20
    dendrite formation, 11, 40
    discharge characteristics, 127–29
    heat generation models, 178–79
    history of development, 13

intercalation materials, 10–11, 40
overview, 10–13
rocking-chair design, 11–12
self-discharge rate, 19
solid-state, 12, 19–21
storage, 39–40
Swing system, 13
LiMn$_2$O$_4$, 59; See also Manganese oxides
Li-polymer batteries, 4, 13–14
Lithium intercalated graphitic carbons, 10, 40
Lower cutoff voltage (LCV), 41
Low-maintenance batteries, 6
Lubricants, 25, 130–31
Lucent Technologies round cells, 6

Magnesium alloys, 25, 40
Maintenance, battery, 6, 8, 60–61
Manganese oxides, 10, 13, 21, 59
Maximum temperature cut-off method, 78–79
MCFC (molten carbonate fuel cells), 16
M&E assembly, 199–200
Meltdowns, 92
Membrane-electrode-assembly (MEA), 17, 195, 199–200
Memory effect, 145, 147; See also Dendrite formation
Methanol, in fuel cells, 192, 196
Mid-point voltage (MPV), 39, 56, 121–22, 126
Models
  air-flow, 138–40
  ambient temperature, 139
  Battery Performance Management System, 135–36
  charge acceptance rates, 175
  depth of discharge, 175, 188
  driveshaft power, 181–83
  electric vehicle performance, 181
  heat generation, 138–39, 177–81
  heat transfer, 138–39, 178–81
  NiMH batteries, 174–76
  polarization resistance, 176
  thermo-electrochemical, 176–88
  vehicle operation, 139
Modular AGM batteries, 8
Module specifications, 42
Molten carbonate fuel cells (MCFC), 16
Motors, power and torque of, 4, 66–67, 183–84
MPV (mid-point voltage), 39, 56, 121–22, 126

National Electrical Code (NEC), 88, 90
National Electric Vehicle Infrastructure Working Council (IWC), 90, 159
National Highway Transportation and Safety Association (NHTSA), 188
Negative temperature coefficient (NTC) thermistors, 76–77, 142
Nickel cadmium (Ni-Cd) batteries, 8–9, 26; See also NiMH batteries
Nickel oxides, 10
NiMH batteries
  activation and formation, 29–30
  advantages, 19
  capacity, 39, 54–57, 123–27
  cell pressure, 175
  charge acceptance, 175
  charge protectors, 142–44
  charging, 74, 78–79, 81–87
  C ratings, 119–20
  current density, 26–27
  depth of discharge, 175
  discharge termination, 124–27
  discharge voltage profiles, 56–58, 121–26
  electrode oxidation, 27–28
  fast charging, 84
  heat capacity, 75
  heat generation rate modeling, 162, 178–81
  hydrogen storage, 9, 39–40
  intelligent chargers, 85–87
  manufacturing process, 9–10, 21
  mathematical model, 174–75
  maximum charge temperature, 78–79
  memory effect, 145, 147
  overcharging, 78–82, 85, 180–81
  oxygen generation, 174–75
  polarization resistance model, 176
  resistance during discharge, 64
  self-discharge rate, 19
  slow charging, 82–84
  state of charge, 175
  storage, 37–39
  temperature effects, 29, 56, 58, 77–79, 123–24, 176
  termination of discharge, 124–27
  vehicle performance test summaries, 211–13
Nissan Motor Company, 5
NTC thermistors, 76–77, 142

Ohmic drop (IR), 99
Open circuit voltage (OCV), 11, 16, 46–47
Operating life tests, 158

Overcharging
    charge protectors, 142, 153
    charge termination methods, 78–80
    fast charging, 96–97, 107–9
    Li-ion batteries, 11
    NiMH batteries, 78–82, 85, 180–81
    VRLA batteries, 69, 71, 73
Over-discharging, 118, 125–26, 153
Overgassing, 73, 109
Overnight charging, 82–84, 110–11
Oversizing, 62
Oxygen
    in NiMH cells, 174–75, 180–81
    release during charging, 80–81
    in VRLA batteries, 33

PAFC (phosphoric acid based fuel cells),
    15–16, 53, 199
Parallel connections, 41, 120
Partial discharge test, 165
Partial oxidation (POX) reactors, 18, 196–97
Partnership of New Generation of Vehicles,
    203
Passivation layers, 53–54, 59
Paxton and Newman model, 174–75
Pb-acid batteries; See Lead-acid batteries;
    Valve regulated lead-acid batteries
Peak load hours, charging during, 110–11
Peak power test, 164
Peak voltage detect (PVD), 86–87
PEFC (polymer electrolyte fuel cells), 195,
    199–201
PEMFC (proton exchange membrane fuel
    cells), 15, 17, 194–95, 198–99
Perfluorocarbon sulfonic acid, 200
Performance, battery, 60–62, 133–59; See
    also Battery Performance Management
    System
Performance safety and abuse test, 167
Performance test summary, 211–13
Peukert relationship, 44, 135, 188
Phosphoric acid based fuel cells (PAFC),
    15–16, 53, 199
Pilot circuit disconnects, 149–50
Pilot line, 157
Platinum catalysts, 15, 17, 192, 195, 197
Polarity reversals, 61, 125–26
Polarization resistance model, 176
Pollution, 1, 14, 187, 191, 193, 201–2
Polyethylene oxide, 14
Polymer electrolyte fuel cells (PEFC), 195,
    199–201

Polymer positive temperature coefficient
    (PPTC) thermistors, 142–45
Polypropylene, 151
Polyvinylidene fluoride, 151
Porosity, 53–54, 173
Potassium hydroxide, 16
Potential differences; See Voltage
Power
    constant power test, 164
    cut-off voltages and, 124–25
    density, 17, 20, 44
    driveshaft, 181–83
    DST power profiles, 115–17
    from the engine, 68
    fuel cell requirements, 198
    generator, 186–87
    inertia, 182
    inverter/system controller, 184
    maximum discharge, 50
    maximum recharge, 50
    motor, 183
    peak power test, 164
    sustained hill-climb power test, 165–66
    traction, 131, 184–85
    variable power discharge test, 164
Power gains, 68
Power losses
    road inclination, 66
    rolling resistance, 65–66
    system controller/engine inefficiency, 67
    transmission inefficiencies, 66–67
    vehicle acceleration, 66
POX (partial oxidation) reactors, 18, 196–97
Preferential oxidation (PROX), 18
Prelyte, 7
Pressure, 74–75, 82, 109, 175
Proton exchange membrane fuel cells
    (PEMFC), 15, 17, 194–95, 198–99
Proton Exchange Technology (PET), 192
PTFE, in fuel cells, 200–201

Ragone plots, 19, 127–28
Range testing, 113–14; See also Driving
    range
RAV4 performance summary, 212–13
Receptacle jacks, 104
Rechargeable batteries, 9
Recombination, 7, 32
Reconditioning charges, 53
Rectifier modules, 104
Regenerative braking, 2–3, 24, 68, 134, 183
Reliability/durability tests, 157, 167–71

Resistance
  average, 49
  discharge and, 64
  installation, 91–92
  instantaneous *vs.* delayed, 55
  NiMH cells, 121
  steady-state, 55
  thermistor, 77, 142–44
Resistive load units, 52, 61
Rest periods, battery, 171–73
Ripple current, 94
Road inclination losses, 66
Rocking-chair design, 11–12
Rolling resistance losses, 65–66, 182

Safety
  battery design, 150–53
  battery requirements, 188–90
  charging equipment, 90–91
  electrical, 151–52
  electric bus isolation, 149
  electric shock, 153–54
  electrolyte spillage, 153–54
  high-voltage disconnects, 149–50
  inductive charging, 111–12
  intrinsic materials hazards, 152–53
  Li-ion batteries, 12
  Li-polymer batteries, 14
  performance safety and abuse test, 167
  water presence, 112
Scaling, 188
Self-discharge rates, 19, 32–33, 35–39, 62,
    96, 165
Separators, 6, 42
Series connections, 41, 71–73, 120
Short circuiting, 152–53
Shuttlecock design, 11–12
Silica gel, 7
Simplified Federal Urban Driving Schedule
    (SFUDS), 123, 165
Six-minute rate, 95
Size, battery pack, 42
Smart batteries, 147–48
SOC; *See* State of charge
Society of Automotive Engineers (SAE), 87,
    90, 148
Solid oxide fuel cells (SOFC), 16, 199
Solid state batteries, 12–14
Sony Corporation, 12
Spare batteries, 62
Special performance test, 165
Specific capacity, 58–59

Specific Discharge Power, 124–25, 128
Specific energy, 44, 59, 173
Specific gravity, 6
Speed, maximum, 185–86
Speedometer calibration, 114, 209
Standloss test, 165
State of charge (SOC)
  benefits of regulation, 47–48
  calculation of, 49–51
  definition, 46, 175
  from open circuit voltage, 46–47
Steam reformer-based fuel processors, 196
Storage, battery, 35–39
Sulfation, 35, 52
Sulphuric acid, 6, 98, 101
Sulphur removal, in fuel cells, 197
Sustained hill-climb power test, 165–66
Switching inverter modules, 104
System controller efficiency, 67–68

Teflon, in fuel cells, 200–201
Telecom applications, 35
Temperature; *See also* Cold weather
  ambient temperature model, 139
  battery capacity and, 44–46, 118–19, 135,
    177
  during charging, 34–35, 70–73, 100
  during discharge, 46, 58, 115–17, 122
  distribution modeling, 176–88
  efficiency, 28–29, 35
  electrolytes, 35, 162
  during fast charging, 97, 100, 108–9
  float charge and, 34
  heat dissipation, 180–81, 197
  heat generation models, 138–39,
    177–81
  heat transfer models, 138–39, 178–81
  midpoint voltage and, 56, 122
  modeling, 138–39, 176–88
  NiMH batteries, 27, 29, 77–79, 123–24,
    176
  nonuniform, 117, 177
  polarization resistance, 176
  reduction in life, 34–35
  sensors, 74–78, 142–45
  solid state Li-ion batteries, 12
  termination methods, 78–80, 83–85
  thermal management system, 137–41
  thermal performance test, 166
  thermal runaway, 92
  VRLA batteries, 26, 28–29, 34–35, 44–46,
    137–38

Terminal posts, 4, 31–32
Thermal capacity, 138
Thermal impedance, 75
Thermal management systems, 137–41, 152, 172–73
Thermal performance test, 166
Thermal runaway, 34, 92, 152, 162, 177–78
Thermistors, 74–78, 142–45
Thermo-electrochemical model, 176–88
Thevenin equivalent circuits, 54, 120
360-second frames, 62–63
Time constants, 55
Tin alloys, 35, 53
TiN_2 alloys, 9
Tires, 130–31, 183
Torque, 4, 66–67, 182–83
Torque converter speed, 66–67
Torque converter torque, 66–67
Toyota Motor Corporation, 5, 212–13
Traction batteries; See Batteries; names of individual battery types
Traction power, 131
Transmission efficiency, 66–67
Trickle charging, 86, 89, 96, 109–10

U. S. Advanced Battery Consortium (USABC), 5, 21, 115, 166
Ultra-fast charging, 86
Urban driving tests, 211–12
Urban driving time power test, 62

Valve regulated lead-acid (VRLA) batteries
    absorbed glass mat, 7–8, 34, 98
    advantages, 5–7, 18–19, 129–30, 151
    capacity definition, 117–19
    capacity discharge testing, 51–53
    capacity recovery, 53–54
    catalysts in, 33–34
    charge protectors, 142
    charging, 34–35, 69–71
    computer simulations, 162
    current density, 26
    cycle life, 44
    discharge tests, 63–64, 116–17
    electrolysis, 33, 108–10
    end of formation, 30–31
    equalization charging, 71
    failure modes, 31–35
    fast charging, 98–99
    formation and EV performance, 23–24
    heat capacity, 101
    hydrogen, 30–31, 33

maximum power calculation, 50
memory effect, 145
overcharging, 69, 71, 73
self-discharge rate, 19
series connections, 71–73
state of charge calculations, 46–51
storage, 36–37
temperature and, 28–29, 34–35, 44–46, 77, 98, 137–38
USABC performance requirements, 43–44
voltage, 60–61
Vanadium oxide, 13
Variable power discharge test, 115, 164
VARTA Li-metal oxide/carbon system, 13
Vehicle acceleration power losses, 66
Vehicle endurance test, 158
Vehicle operational model, 139
Vehicles, electric; See Electric vehicles
Vehicle tire limit, 183
Ventilation, 90–92, 129–30, 140, 152, 159
Vibration test, 166
Voltage
    average, 49
    during charging, 74–75, 99–100
    clamp, 69–71, 73
    cut-off points, 41, 118, 124
    discharge profiles, 55–58, 119–26
    end-of-discharge, 126–27
    fuel cells, 16
    Li-ion batteries, 11
    under load, 119–20, 184–86
    mid-point, 39, 56, 121–22, 126
    minimum, 185–86
    NiMH battery packs, 176
    no-load, 135
    overview, 6–7
    profile calculation, 50–51
    resistance-free, 98, 100
    thermistor, 77, 145
    transition voltage restoration current, 84
    variation during formation cycles, 27
Voltage peak method, 97
VRLA batteries; See Valve regulated lead-acid batteries

Water loss, 32, 34–35
Weak cells, 61, 126, 133–34
Wheel bearing losses, 182
Zero-emission vehicles, 191, 193, 202–3
Zinc oxide catalysts, 197
Zinc precipitation, 109
Zr-oxide, solid doped, 16